T. Mann, J. Banks

A Treatise on Rivers and Canals

By Theod Aug. Mann, Member of the Imperial and Royal Academy of

Sciences at Brussels; Communicated by Joseph Banks, Esq. P. R. S.

T. Mann, J. Banks

A Treatise on Rivers and Canals
By Theod Aug. Mann, Member of the Imperial and Royal Academy of Sciences at Brussels; Communicated by Joseph Banks, Esq. P. R. S.

ISBN/EAN: 9783337346096

Printed in Europe, USA, Canada, Australia, Japan

Cover: Foto ©berggeist007 / pixelio.de

More available books at **www.hansebooks.com**

XXXVII. *A Treatife on Rivers and Canals.* By Theod.
Aug. Mann, *Member of the Imperial and Royal Aca-
demy of Sciences at* Bruffels ; *communicated by* Jofeph
Banks, *Efq. P. R. S.*

Read June 24, 1779.

T O   J O S E P H   B A N K S,   E S Q,   P. R. S:

S I R,

YOUR election to the Prefidency of the firft literary
and fcientific Society in the world, to a chair fo
long and fo glorioufly occupied by the great NEWTON;
joined to the friendfhip you have been pleafed to honour
me with fince my being firft known to you: has encou-
raged me to fend you fomething of my compofition, as
the beft way of expreffing my fincere refpect and attachment to you, and my profound veneration for the illuftrious Body which has chofen you for its head. Though
various circumftances, by carrying me very early into foreign countries, have made me from my youth almoft
an

an alien to my native foil, and put me in a fituatio. which apparently muft make me ever remain fo; yet, neither time nor diftance could ever weaken, much lefs obliterate, my tender attachment to it, or my ardent wifhes for its welfare.

Thefe confiderations will, I hope, merit a favourable acceptance from the Royal Society of the following piece, which I have the honour of addreffing to you; and an indulgent condefcenfion for its imperfection in every re- fpect, and particularly in point of ftyle. Five and twenty years abfence from my native country, and the neceffity of converfing during that time in different foreign lan- guages, muft unavoidably have filled mine, without my being fenfible of it, with idioms and expreffions in no wife Englifh.

As to the fubject I have undertaken to treat on this occafion, I was guided in the choice thereof by the mo- tive of faying fomething that might be ufeful to my na- tive country. The great number of extenfive and mag- nificent canals, which have been cut through almoft every part of England of late years, for the ufe of inter- nal navigation, and which do honour to the public fpirit of the nation, merit to be confidered in a fcientifical as well as in a commercial light. Their waters have their laws of motion different in many cafes from thofe of

7   rivers:

rivers: they are liable to many accidents which the others are not, and of a different nature. Thefe accidents do not become fenfible till many years after their conftruction, and are better prevented in time than remedied when they happen. I have long lived in a country famous for its navigable canals, and have been much employed, under the eyes of the government of it, upon that fubject. I only mention this to fhew, that I have not undertaken to treat a fubject to which I am an utter ftranger.

There are, moreover, many confiderations concerning the laws of motion in rivers and canals in general, the velocity of their currents in proportion to the quantity of their declivity, and the means of afcertaining therefrom the refpective heights of the interior parts of continents, which merit the attention of a natural philofopher. I fhall venture to offer my thoughts and obfervations (fome of which, I believe, are new) on all thefe fubjects in the enfuing Differtation, which I fubmit entirely to the judgment of the Royal Society, and fhall efteem myfelf happy if I fucceed in it, fo as to be of any ufe to my country, and to be able to teftify, at the fame time, my profound refpect and veneration to you, DEAR SIR, and to the illuftrious body over which you prefide.

SECTION I.

Different ufes for which canals are made, with an account of the principal authors who have wrote concerning them.

1. Artificial canals are to be confidered in a double light; as facilitating commerce by means of internal navigation, and as preventing inundations by carrying off the too great abundance of water from low and flat countries, fuch as are Holland, Flanders, &c. In thefe laft named countries they ferve at once for both purpofes; and it is in this double light that I fhall confider them in the enfuing difcourfe. If canals for draining have fluices upon them, particularly at the end whereby they difcharge their waters, as is univerfally the cafe in the Low Countries, they differ in no wife from navigable canals: if they have nothing to fuftain their waters in them, they are to be confidered in every refpect as rivers or rivulets, and follow the fame laws. It muft, therefore, be carefully kept in mind, that whenever I mention canals, I mean thofe only whofe waters are kept up by fluices, and never thofe without them, which I include, without diftinction, under the common appellation of rivers; for

they

they are no more than artificial ones; If I miftake not, all the navigable canals in England are of the firft fort; that is, have their waters kept up, and let off by fluices. This neceffary diftinction will take away all ambiguity from what I have to fay on canals throughout the following difcourfe.

2. But that I may fulfil the tafk I have undertaken, it is neceffary firft of all to lay down fuch principles on the nature of rivers and canals in general as have been demonftrated true both by calculation and experience; to the end, that we may deduce from thence the true laws of motion of their waters, and the quantity of declivity of their beds: for this purpofe, and becaufe a large volume would hardly fuffice to comprife all the demonftrations of thefe principles, which, confequently, I am obliged to omit in this treatife, it will not be amifs to mention the principal authors who have treated this fubject in different ages and countries, in whofe works the demonftrations of all the principles I fhall lay down may be found, if any one doubts the truth of them. Thefe are the following.

SEXTUS JULIUS FRONTINUS, de Aquæ-ductibus Urbis Romæ, cum Notis POLENI, impreff. 1722.

JOHN BAPTIST ALEOTTI, Hydrometrician to the Duke of Ferrara, and to Pope CLEMENT the VIIIth.

* Don

* Don BENEDICT CASTELLI, Benedictine Abbot, de Menfurâ Aquarum Currentium.

J. B. BARATTERI, de Architettura d'Acque, lib. VI. Piacenza, in folio, 1656.

ALEXANDER BELTINZOLI, of Cremona.

NICOLAUS CABEUS, in Libris Meteorum.

GALILEI GALILEO.

* JOH. BAPT. BALIANI, de Motu Liquidorum.

* JOH. BAPT. RICCIOLI, Geographiæ et Hydrographiæ Reform. libro VI. cc. 29. et 30.

* CLAUDE MILLET DESCHALES, de Fontibus et Fluminibus, à prop. 39. ufque ad 56.

VARENNIUS, General Geography, with Dr. JURIN's and Dr. SHAW's Notes, edit. of 1765, vol. I. from page 295 to page 358.

Dr. JURIN, in the Philofophical Tranfactions, N° 355. page 748. et feq.

MARIOTTE, Traité du Mouvement des Eaux.

VARIGNON, Memoires de l'Academie des Sciences de Paris, pour 1699 et 1703.

* Sir ISAAC NEWTON, Princip. Mathem. lib. II. § 7. page 318 et feq. edit. 1726.

* DANIELIS BERNOUILLI, Hydrodynamica, in quarto, Argentorati, 1738.

3   * DOME-

* DOMENICHE GUGLIELMINI, della Natura de Fiumi, Bononiæ, 1697, in quarto. Ejuſdem de Menſurâ Aquarum fluentium, Bononiæ, in quarto.

* JOH. POLENUS, de Caſtellis et de Motu Aquæ mixto, Patavii, 1697, 1718, 1723.

* RACCOLTA d'Autori che trattano del Moto dell' Acque, Fiorenza, 1723, 3 vol. quarto, cum fig.

JAC. HERMANNUS, in Phoronomia, cap 10. page 226. et ſeq.

CHRIST. WOLF, Curſ. Mathem. Hydraulicæ, cap. VI. edit. Genevæ, in quarto, 1740.

M. DE BUFFON, ſur les Fleuves, dans ſon Hiſtoire Naturelle, tom. II. p. 38—100. de la 1ᵉʳᵉ edit. en 12mo.

Several Memoirs upon this Subject in the Collection of the Royal Academy of Sciences of Paris, particularly thoſe of M. PITOT, in the volumes for 1730 and 1732.

S'GRAVESANDE, in Elementis Phyſicæ, tom. I. lib. II. cap. 10.

* R. P. LECCHI S. J. Hydroſtatica, Mediolani, 1765. In this excellent work are ſeveral pieces by Father BOSCOVICH upon the ſame ſubject.

* STATTLERI Phyſica, vol. III. p. 232—286. de curſu Fluminum, ejuſque Menſuratione et Directione. Aug. Vindel. 1772, 8 vol. in octavo. This author gives many late obſervations and experiments on the motion and meaſure

measure of currents, as those of ZENDRINI, HIMENII, &c.

Two other authors have lately wrote upon rivers and canals, but their works are not yet come to my hands; to wit,

Father FRISI, an Italian Barnabite, Professor of Mathematics at Milan.

M. DE LA LANDE, of the Royal Academy of Sciences at Paris, who has just published a History in folio, with plates, of all the Navigable Canals in the World that have come to his Knowledge.

Among the above authors, those marked with an asterisk (*) are they who have treated the subject in question with the greatest exactness or most extent; and it is from them chiefly that I shall lay down such principles and laws of action in rivers and canals as regard the subject I have taken in hand. By this means I shall avoid advancing any thing upon so important a matter, but what is founded upon the most certain and exact experiments, and conformable to what are demonstrated to be the real and unalterable laws of nature.

SEC-

SECTION II.

The theory of rivers and canals.

I. DEFINITIONS.

3. A river is a greater or leſſer quantity of water which runs conſtantly, by its own gravity, from the more elevated parts of the earth, towards thoſe which are more depreſſed, in a natural bed or channel open above.

4. If this bed or channel is artificial, and has been dug by hands, it is called *a canal,* of which there are two kinds; thoſe where the channel is every where open, and without ſluices, which I call an *artificial river*; and thoſe where the waters are kept up or let off by the means of ſluices: it is this ſecond ſort which I ſhall call hence-forwards by the proper name of *a canal.*

5. *A river is ſaid to perſevere in the ſame ſtate* ſo long as there runs off an equal quantity of water in the ſame time, without any increaſe or diminution, ſo that it remains always at the ſame height in the ſame place. When the circumſtances are different from this, it is ſaid reſpectively that *a river increaſes or diminiſhes.*

6. *A ſection* of the bed of a river or canal is a plane drawn perpendicular to the bottom of the bed and to the

2   direction

direction of the ftream of water, and whofe limits are thofe of the water itfelf which runs off in that place.

7. I call *fections of equal velocity*, all thofe where the water runs with equal velocity; and *fections of greater or leffer velocity*, thofe where the water runs fafter or flower refpectively, and when compared to others.

8. I call *mean velocity of a current or ftream of water*, that which a river or canal would have, if all the parts thereof, to wit, thofe of the bottom, the fides, the middle, and the furface of the fame fection, ran with an equal velocity, in fuch a manner that there would pafs juft as much water in the fame time by this uniform motion, as there does now actually pafs by the irregular flowing of the ftream.

II. PROPOSITIONS, *or laws of action in rivers and canals.*

9. The motion of water in rivers proceeds from the fame principle which produces the defcent of heavy bodies upon inclined planes.

11. The defcent of heavy bodies upon inclined planes follows exactly the fame laws as thofe obferved in the defcent of heavy bodies in a perpendicular line towards the center of the earth; that is,

1ft, They defcend by a motion uniformly accelerated.

2dly, The

2dly, The fpaces, run over by heavy bodies which fall perpendicularly by a motion uniformly accelerated, are in *a duplicate ratio of the times and velocities* refpectively.

3dly, Thefe fpaces, in equal times, augment in the fame ratio as the odd numbers in progreffion 1, 3, 5, 7, 9, 11, &c.

4thly, Therefore, both the times and the velocities are in a *fub-duplicate ratio of the fpaces run over*.

5thly, It is demonftrated by the principles of mechanics, that the velocity acquired by a heavy body defcending freely upon an inclined plane, in a given time, is to the velocity which the body would acquire in the fame time, by falling perpendicularly, as the height of the inclined plane is to its length.

6thly, From whence it follows, that the velocities which bodies acquire in their defcent upon inclined planes, are *in a direct ratio of the fquare roots of the quantity of inclination or declivity of the planes*.

11. So that when water flows freely upon an inclined bed, it acquires a velocity, which is always as the fquare root of the quantity of declivity of the bed.

12. In an horizontal bed, opened by fluices or otherwife, at one or both ends, the water flows out by its gravity alone; and the flowing is quicker or flower in a di-

rect

rect ratio of the refpective heights of the water, by reafon of the weight of the fuperior waters upon the inferior.

13. From hence (N° 11, 12.) it follows, firft, that as much as the declivity of the bed or channel of a river is greater, fo much alfo will the velocity of the flowing waters be proportionably increafed.

2dly, As much as the water in an horizontal bed is deeper, fo much will the velocity of the current be increafed; and this velocity will diminifh in proportion to the decreafing depths of the water in the bed.

3dly, Abftracting from the refiftance caufed by the bottom and fides of the bed, as much nearer as the water is to the bottom, fo much will its motion be accelerated; not only becaufe the inferior waters are more compreffed by the fuperior in proportion to their greater depth; but alfo becaufe the inferior ones have a greater declivity than the fuperior, by reafon of their greater depth in the bed, where they are more depreffed with refpect to the elevation of their common fource or fpring. But thefe different velocities of the upper and lower waters in the fame fection of the bed (abftracting from the friction of the bottom and fides) approximate indefinitely to each other in proportion to the length of the channel, but ftill without a poffibility of their ever becoming equal in fact, if they met with no refiftance from the bed.

14. There-

14. Therefore, the motion of water flowing freely in an inclined channel, is accelerated by its own weight combined with the quantity of declivity in the bed.

Nevertheleſs, the velocity of waters which flow in an inclined bed, during their actual flowing, is not accelerated by the weight which the inferior waters ſuſtain from the ſuperior ones, in caſe the lower parts have already, by the declivity of the bed, a greater velocity than that which the weight of the ſuperior ones impreſſes upon them. The reaſon of which is, that no body which follows another with a leſſer velocity can act by impulſion upon that which precedes it with a greater velocity, as is the caſe with regard to theſe ſuperior and inferior waters. But the weight of the upper waters begins to accelerate the lower as ſoon as they fall into an horizontal bed, or one that is ſo nearly horizontal as to deſtroy the greater velocity of the lower waters above that of the upper.

15. The velocity of rivers depends ſometimes upon the ſole declivity of their beds; ſometimes alſo upon the ſole gravity of their waters: and if theſe two cauſes ſometimes act together, the effect produced is only the reſpective exceſs of the one above the other. It often happens in the ſame ſection of a river, that the acceleration of velocity in the inferior parts proceeds from the weight of

4 E 2   the

the upper waters, while that in the upper parts proceeds from the declivity of the bed.

From whence it follows, that in rivers which have little declivity, it is the depth of the waters which contributes moft to accelerate their current; and in thofe whofe beds have moft declivity, it is the defcent of gravity upon an inclined plane which has the greateft fhare. in producing this acceleration.

To find whether the water in a part of a river where the bed is nearly horizontal flows by the velocity acquired in the preceding declivities, or by the compreffion of the upper waters upon the lower in that place; a pole muft be thruft down to the bottom, and held perpendicular to the current of the water, with its upper end above the furface: if the water fwells and rifes immediately againft the pole, it fhews that its flowing is by virtue of a preceding declivity: if, on the contrary, the water ftops for fome moments before it begins to rife againft the pole, it is a proof that it flows by means of the compreffion of the upper waters upon the lower.

16. The *abfolute height or elevation* of the furface of a river which perfeveres in the fame ftate (N° 5.) continually decreafes, as the diftance in the river from its fource increafes; by reafon that its bed muft continually, incline and tend towards the center of the earth.

17. The

17. The velocity of each particle of water in a regular channel, that is, where the bed is a regularly inclined plane, may be determined by drawing a perpendicular from the particle propofed to the horizontal curve which paffes through the fpring, or that point of the river where the particle in queftion begins to acquire its velocity. For the velocity which this particle would acquire, in falling freely along the faid perpendicular, is the fame as that which it has acquired in its defcent along the inclined plane of its bed.

18. So long as *a river perfeveres in the fame ftate.* (N° 5.) there flows an equal quantity of water in equal times, how unequal foever the fections be through which they flow; and, confequently, where the fection of the river is greater, the velocity of the flowing water is lefs; and where the fection is lefs, the velocity is greater; always in an inverfe proportion. From hence may be deduced the following and other fimilar propofitions.

1ft, Through equal fections, in equal times, and with equal velocity, there muft flow equal quantities of water.

2dly, Through equal fections, in equal times, but with unequal velocities, the quantities of water which flow, are in a direct ratio of the refpective velocities.

3dly, Through

3dly, Through unequal fections, in equal times, and
with equal velocities, the quantities of water which pafs
fire in a direct ratio of their refpective fections.

4thly, Through unequal fections, and with unequal
velocities, the quantities of water which flow in equal
times, are in a combined ratio of the fections and mean
velocities (N° 8.) together.

In a word, the fections of the bed, the mean velocities, the times of flowing, and the quantity which flows,
are univerfally in a combined ratio together; and this
combination is what is called *the momentum of a river*;
and this *momentum of the fame flowing water is univerfally equal.*

19. From hence may eafily be deduced the principles
for calculating the quantity of diminution of the water
in a lake, pond, or veffel, by any determinate flowings
whatfoever: for as the furface of the lake, &c. is to the
fection of the current which carries off the waters; fo is
the mean velocity of the current in this fection to the
decreafe of the waters in the lake, &c. and *vice verfâ.*

III. *On the nature of rivers and flowing waters.*

20. Rivers contain divers inherent caufes of the acceleration of their motion.

1ft, Their fprings are either in mountains or on high grounds, and it is by the defcent of the waters from thefe elevations that they acquire a velocity and acceleration of motion fufficient to fuftain and propagate it through the reft of their courfe..

2dly, The cohefion of the particles of a fluid, *in a bed ever fo little inclined,* is a fecond caufe of acceleration of motion in the fluid; becaufe, by their mutual attraction, thofe particles which begin firft to flow draw after them thofe which are contiguous, thefe the following, and fo on *ad infinitum.*

3dly, Moreover, where a river, by flowing in a bed nearly horizontal, has loft a great part of the velocity which it had acquired in the preceding declivities, and the bed by this means is become large and fhallow, which confequently again augments the flownefs of the current; it may, however, recover a part of its velocity, even in the fame horizontal bed, by augmenting the depth thereof, and diminifhing its breadth; for by this means the weight of the fuperior waters upon the infe-

7   rior

rior is increafed, and confequently the velocity of the whole is augmented (N° 13. 18.). In the fame manner a junction of rivers in the fame bed, by excavating and deepening it, augment the velocity of the common cur-rent, as we fhall fhew more particularly hereafter.

21. On the other hand, flowing waters meet with many powerful caufes of refiftance to their motion, which tend continually to diminifh their velocity. Such are the following:

1ft, The attraction and continual friction of the bot-tom and the fides of the bed, contribute greatly towards retarding the motion of the water.

2dly, The fame effect is produced likewife by the many obftacles which they meet with in their way; fuch as inequalities in the bottom and fides of the channel, banks of fand and mud, rocks, trunks of trees, and other fuch things.

3dly, The many windings and angles made in their courfe, which produce fo much the more refiftance and hindrance to the motion of the water, as the courfe varies more and oftener from a right-line.

4thly, The diminution of their declivity the farther they recede from their fprings; this being generally the leaft towards their mouths, which are for the moft part in extenfive plains.

Finally,

Finally, the natural cohefion of the particles of water *in an horizontal bed* contributes to retard its motion precifely by the fame force which contributes to accelerate it *in an inclined bed.* By diminifhing or taking away the above obftacles to the free motion of water in rivers and canals, the velocity of their currents will be increafed in the fame proportion, and thereby alfo all the dangers and ravages of inundations may be prevented, as we fhall fhew hereafter.

But if fome or all of thefe caufes, in a greater or leffer degree, did not exift in rivers of confiderable depth and declivity, it is demonftrated [a], that the velocity of their currents would be accelerated to twelve, fifteen, and, in fome cafes, even to twenty times more than it is at prefent in the fame rivers, whereby they would become ab- abfolutely unnavigable.

22. The waters in a river or open canal have their motion accelerated, fo long as the effects proceeding from gravitation, declivity, depth, in a word, fo long as *the fum of accelerations furpaffes the fum of refiftances.*

When thefe different fums become equal to each other, the motion of the water is neither accelerated nor retarded, but remains equal, till fomething anew deftroys the *equilibrium.*

[a] Vide LECCHI, Hydroftat. et STATTLERI, Phyfic. tom. III. p. 252.

When the fum of refiftances and caufes of retardation is greater than the fum of accelerating caufes, the velocity of the river is diminifhed in proportion to the excefs.

23. The *percuffion* of the waters of a river againft an obftacle which is oppofed to their motion, is the action of the waters ftriking againft that obftacle; and the principles for calculating the quantity of this percuffion, or the effects which any obftacles whatever produce in the motion of rivers, by known forces, and in determinate times, are as follows:

1ft, The percuffion of the water of a river againft any obftacle whatever is univerfally in a compound ratio of the quantity of the plane or planes which the obftacle oppofes to the current of water, of the fine of the angle of incidence which the direction of the current makes with thefe planes, and of the fquare of the velocity of the faid current.

2dly, The refiftance, therefore, which the bed of a river oppofes to its current from any particular obftacles in it, is in a compound ratio of the magnitude and fituation of the planes of thefe obftacles, together with the fquare of the current's velocity in the place where thofe obftacles are found.

3dly, The accelerating force of a river, or that by which it furmounts the refiftance of its bed in any one

7   place,

place, compared to that in another, is in a compound ratio of the mafs of water and of the velocity of the current in thofe places refpectively.

24. In different parts of the fame river, the velocity of the current is greater in a direct proportion of the greater declivity of the bed; becaufe the relative gravity of the flowing particles augments in that ratio.

25. But in the fame fection of a river, the fuperior parts, and thofe which are fartheft from the bottom and the fides, will continue their courfe by the fole caufe of the declivity of their bed, *how little foever it be*; becaufe thefe waters not being retarded by the friction of the bottom and fides of the bed, or hardly by any other ob-ftacle whatever, *the leaft poffible deviation from a level will produce a current.* But the waters at the bottom of a river, both becaufe of their friction againft it, and of the irregularities which are almoft every where found in it, will lofe that little motion which a very fmall declivity can give them, and their motion in that cafe will be pro-duced alone by the compreffion of the fuperior waters upon them.

The inferior waters which thus acquire their motion from the weight of the fuperior ones upon them, com-municate reciprocally a part of their motion, by means of the natural cohefion of the particles together (N° 20.)

to

to the fuperior ones, which in an horizontal bed, without this caufe, would have no other motion than that which is impreffed upon them by the impulfive force of the waters defcending from their elevated fprings.

From whence it appears, that the fuperior and inferior waters in a river communicate reciprocally a part of their motion to each other; but this can never go beyond a certain point or *maximum,* which is always proportionable to the *momentum* of the river in that place (N° 18.).

It follows from hence, that the greateft velocity of a river, running in a right-line, is in *the center of its fection* (N° 6.); that is to fay, in that point which is the fartheft poffible from the furface of the water and from the bottom and fides of the bed, all taken together. This part has the advantage of one half of the depth of water preffing upon it, and it is exempt from the friction of the bottom and fides of the bed which are there overcome and vanifh by the perpendicular compreffion.

On the contrary, the leaft velocity of the water is at the bottom and fides of the bed, becaufe it is there that the refiftance produced by friction is greateft, from whence it is communicated to the other parts of the fection in *an inverfe duplicate proportion of the diftances from the bottom and fides combined together,* until it becomes a
negative

negative quantity, where the effect vanishes, or is reduced to nothing.

26. The best and most simple method of measuring the velocity of the current of a river or open canal, that I know of, is the following:

Take a cylindrical piece of dry, light wood, and of a length something less than the depth of the water in the river: round one end of it let there be suspended as many small weights as may be necessary to keep up the cylinder in a perpendicular situation in the water, and in such a manner that the other end of it may just appear above the surface of the water. Fix to the center of that end which appears above water a small and straight rod, precisely in the direction of the cylinder's axis; to the end, that when the instrument is suspended in the water, the deviations of the rod from a perpendicularity to the surface of it may indicate which end of the cylinder advances the fastest, whereby may be discovered the different velocities of the water at different depths; for if the rod inclines forwards according to the direction of the current, it is a proof that the surface of the water has the greatest velocity; but if it inclines back, it shews that the swiftest current is at the bottom; if it remains perpendicular, it is a sign that the velocities at the surface and bottom are equal.

This

578   *Mr.* MANN's *Treatife*

This inftrument being placed in the current of a river or canal receives all the percuffions of the water throughout the whole depth, and will have an equal velocity with that of the whole current *from the furface to the bottom* at the place where it is put in, and by that means may be found, both with eafe and exactnefs, the mean velocity of that part of the river for any determinate diftance and time.

But to obtain the mean velocity of the whole fection of the river, the inftrument muft be put fucceffively both in the middle and towards the fides, becaufe the velocities at thofe places are often very different from each other. Having by this means found the *difference of time required for the currents to run over an equal fpace;* or, *the different diftances run over in equal times, the mean proportional* of all thefe trials, which is found by dividing the common fum of them all by the number of trials, *will be the mean velocity of the river or canal.*

If it be required to find the velocity of the current only at the furface, or at the middle, or at the bottom, a fphere of wood, of fuch a weight as will remain fufpended in equilibrium with the water at the furface or depth which we want to meafure, will be better for the purpofe than a cylinder, becaufe it is only affected by the
water

water of that fole part of the current where it remains fufpended.

It is very eafy to guide both the cylinder and the globe in that part which we want to meafure, by means of two threads or fmall cords, which two perfons muft hold and direct, one on each fide the river; taking care at the fame time neither to retard nor accelerate the motion of the inftrument.

Several other methods have been invented for determining the velocity of the currents of rivers and canals, which may be feen in moft of the authors enumerated in the beginning of this effay (N° 2.)

IV. *Application of the preceding laws of the acceleration and retardation of currents to rivers and canals in general, from whence are deduced the various means of preventing or remedying the defects and inconveniencies which muft neceffarily happen to them in a feries of years.*

27. By combining together all we have faid hitherto upon the nature and theory of motion in rivers, and particularly in the articles 13. 18. 20. 21. and 23 it follows evidently, that *the deeper the waters are in their bed in proportion to its breadth, the more their motion is accelerated;*

celerated; fo that *their velocity increafes in an inverfe ratio of the breadth of the bed, and alfo of the greatnefs of the section*; from whence are deduced the two following univerfal practical rules:

1ft, To augment the velocity of water in a river or canal, without augmenting the declivity of the bed, we muft *increafe the depth and diminifh the breadth of its bed.*

2dly, But to diminifh the velocity of water in a river or canal, we muft, on the contrary, *increafe the breadth and diminifh the depth of its bed.*

The above propofition is perfectly conformable to obfervation and experience; for it is conftantly feen, that the current is the fwifteft where the waters are deepeft and the breadth of the bed the leaft; and that they flow floweft where their depth is the leaft and the breadth of the bed the greateft.   " The velocity of waters," fays M. DE BUFFON [b], " augments in the fame proportion as " the fection of the channel through which they pafs " diminifhes, *the force of impulfion from the back-waters* " *being fuppofed always the fame.*   Nothing," continues he, " produces fo great a diminution in the fwiftnefs of " a current as its growing fhallow; and, on the contrary, " the increafe of the volume of water augments its

[b] Hift. Nat. tom. II. p. 53. 60. edit. in 12mo.
5   " velocity

" velocity more than any other caufe whatever. The celebrated WOLF, in his Hydraulics [a], affures us, that " it is a conftant and univerfal practice, for accelerating " the current of waters, to deepen the bed, and at the " fame time to render it narrower."

28. When the velocity which a river has acquired by the elevation of its fprings and the impulfe of the back-water, is at laft totally deftroyed by the different caufes of refiftance which we have enumerated above (N° 21.) becoming equal or greater than the firft, the bed and current at the fame time being exactly horizontal, nothing elfe remains to propagate the motion, except *the fole perpendicular compreffion of the upper waters upon the lower, which is always in a direct ratio of their depth.* But this neceffary refource, this remaining caufe of motion in rivers, augments in proportion as all the other diminifh, and as the want of it increafes: for as the waters of rivers in extenfive plains lofe the acceleration of motion acquired in their defcent from their fprings, their quantity accumulates in the fame bed by the junction of feveral ftreams together, and their depth increafes in confequence thereof. This junction and fucceffive accumulation of many ftreams in the fame bed, which we fee univerfally in a greater or lefs degree in all rivers throughout the known world, and which is fo abfolutely

(c) N. 224.

neceffary

neceffary to the motion of their waters, can only be attributed, fays Signor GUGLIELMINI [d], to the infinite wifdom of the fupreme Author of Nature.

29. The velocity of flowing waters is very far from
being in proporcion to the quantity of declivity in their
bed: if it was, a river whofe declivity is uniform and
double to that of another, ought only to run with double the fwiftnefs when compared to it; but in effect it is
found to have a much greater, and its rapidity, inftead of
being only double, will be triple, quadruple, and fometimes even more: for its velocity depends much more on
the quantity and depth of the water, and on the compreffion of the upper waters on the lower, than on
the declivity of the bed. Confequently, whenever the
bed of a river or canal is to be dug, the declivity muft
not be diftributed equally throughout the whole length;
but, to give a fwifter current to the water, the declivity
muft be made much greater in the beginning of its
courfe than towards the end where it difembogues itfelf,
and where the declivity muft be almoft infenfible, as we
fee is the cafe in all natural rivers; for when they approach near the fea their declivity is little or nothing,
yet they flow with a rapidity which is fo much greater,
as they contain a greater volume of water: fo that in

(d) Della Natura de Fiumi.

great

great rivers, although a large extent of their bed next the sea should be absolutely horizontal, and without any declivity at all, yet their waters do not cease to flow, and to flow even with great rapidity, both from the impulsion of the back waters, and from the compression of the upper waters upon the lower in the same section.

30. Whoever is well acquainted with the principles of the higher geometry, will easily perceive that it would be no difficult matter so to dig the bed of a canal or river, that *the velocity of the current should be every where equal.* It would be only giving it the form of a curve along which a moving body should recede from a given point, and *describe spaces every where proportional to the times,* allowance being made therein for the quantity of effect of the compression of the upper waters upon the lower. This curve is what is called the *Horizontal Isochronic,* being the flattest of an infinity of others which would equally answer the problem where fluids were not concerned. Upon these curves may be seen LEIBNITZ, HUYGHENS, and the two BERNOUILLI's, who were the first that determined and analysed them, and also many succeeding geometricians, if any one is desirous to occupy himself in such speculations as are more curious than useful, which is not my purpose in this treatise.

31. Not-

31. Notwithftanding all we have faid concerning the neceffity of augmenting the depth of a river in a greater proportion than its breadth, if we would accelerate its current; yet it is certain, that this can only be done to a certain point, without deftroying that equilibrium which ought to reign between the depth and the breadth of the fection of the ftream, and thereby putting the river into a ftate of continual violence, which will inceffantly exert itfelf to the deftruction of the banks and wiers made to keep it in, and that action will always exert itfelf in a direct ratio of the greater or lefs want of equilibrium, as it would be eafy to demonftrate by the principles of hydraulics. Thefe fame principles give likewife the juft proportions of this equilibrium between the perpendicular and lateral compreffion of the water in any river or canal whatfoever, which vary in an inverfe proportion, according to the different degrees of the declivity and velocity of the current; and in a direct one of the greater or lefs coherence and hardnefs of the fubftances which compofe the bed. Rivers which flow in beds compofed of homogeneous matter of little confiftency, fuch as fand, &c. are always more broad than deep, when compared to thofe which run in beds of matter of greater tenacity. It is manifeft, that the equilibrium here fpoken of is real,

4   becaufe

becaufe rivers remaining in the fame ftate only widen
their beds to a certain pitch which they do not furpafs.

32. M. DE BUFFON remarks, " That people accuf-
" tomed to rivers can eafily foretell when there is going
" to be a fudden increafe of water in the bed from floods
" produced by fudden falls of rain in the higher countries.
" through which the rivers pafs. This they perceive by
" a particular motion in the water, which they exprefs
" in their dialect, by faying that *the river's bottom moves*;
" that is, the water at the bottom of a channel runs off
" fafter than ufual; and this increafe of motion at the
" bottom of the river always announces a fudden in-
" creafe of water coming down the ftream. Nor does
" their opinion therein," continues the fame author,
" feem to be ill-grounded on the nature of things; for
" the motion and weight of the waters coming down,
" though not yet arrived, muft act upon the waters in
" the lower parts of the river, and communicate by im-
" pulfion part of their motion thereto; fince a canal or
" river contained in its bed is to be confidered in fome
" degree as a column of water contained in a long tube,
" where the motion is communicated at once throughout
" the whole length." In a river or canal, open above, it
is only communicated to a certain diftance; that is, as far
as the impulfive force of the new increafe and fuperior
rapidity

rapidity of the back-waters acts upon the ftream, which will always be as far as till this force is gradually, and at laft wholly, deftroyed by the fuperior gravitation of the fuper-incumbent waters in the ftream. Something of the fame kind happens when a very great additional weight comes fuddenly upon the furface of a river or canal; for inftance, by the launching of a fhip or of feveral boats together upon it. Thefe caufes increafe the velocity of the water in the lower parts of the bed, and moreover retard its motion at the furface, which effect may properly be called *making the river's bottom move.* For the fame reafon, the increafe of weight of the waters in a fudden flood, as well as the increafe of their impulfive force, muft contribute to produce this effect, and, by increafing the motion in the bottom of the river, may hinder, for fome fpace of time, the ftream from fenfibly rifing in the bed.

33. All obftacles whatever in the bed of a river or canal, fuch as rocks, trunks of trees, banks of fand and mud, &c. muft neceffarily hinder proportionably the free running off of the water; for it is evident, from what we have faid, that the waters fo far back from thefe obftacles, until the horizontal level of the bottom of the bed becomes higher than the top of the obftacles, muft be intirely kept up and hindered from running off in proportion

portion thereto (N° 23.). Now as the waters muft continue to come down from their fources, if their free running off is hindered by any obftacles whatever, their relative height back from them muft neceffarily be increafed
until their elevation, combined with the velocity of their
current proceeding from it, be arrived to fuch a pitch at
the point where the obftacles exift, as to counterbalance
the quantity of oppofition or impediment proceeding
from thence, which frequently does not happen until
all the lower parts of the country round about are laid
under water.

34. Now it is certain from all experience, that the
beds of rivers and canals in general are fubject to fome
or others of the obftacles above mentioned. If rocks or
trees do not bar their channels, at leaft the quantity of
fand, earth, and mud, which their ftreams never fail to
bring down, particularly in floods, and which are un-
equally depofed according to the various windings and
degrees of fwiftnefs in the current, muft unavoidably, in
courfe of time, fill up, in part, different places in the
channel, and thereby hinder the free running off of the
back waters. This is certainly the cafe, more or lefs, in
all rivers, and in all canals of long ftanding, as is
notorious to all thofe well acquainted with them.
Hence, if thefe accidents are not carefully, and with a

conftant

conftant attention prevented, come inundations, which fometimes lay wafte whole diftricts, and ruin the fineft tracts of ground, by covering them with fand: hence rivers become unnavigable, and canals ufelefs, for the purpofes for which they were conftructed. Canals, in particular, by reafon that their waters for the moft part remain ftagnant in them, are ftill more liable than rivers to have their beds fill up by the fubfiding of mud, and that efpecially for fome diftance above each of their fluices; infomuch, that if continual care is not taken to prevent it, or remedy it as often as it happens, they will foon become incapable of receiving and paffing the fame veffels as formerly. Nay, the very fluices themfelves, if the floors of their bottoms are not of a depth conformable to the bed of the canal, will produce the fame accidents as thofe we have been fpeaking of; for if they are placed too low, they will be continually filling up with fand or mud; if too high, they have the fame effect as banks or bars in the bed of a river, that is, they hinder all the back-waters under their level from running off, and foon fill up the bed to that height by the fubfiding of mud. This effect is much accelerated by the fhutting of the lower fluices, which makes a great volume of water reflow back to thofe next above them, till the whole is filled and becomes ftagnant. Now it is evident,

that

that this ftate of things muft contribute far more to the fubfidency of mud and all other matters brought down by the waters in canals, than can be the cafe in rivers whofe currents conftantly flow.

35. I do not fuppofe that thefe inconveniencies can have yet manifefted themfelves by any very fenfible effects in the many new canals and fluices lately conftructed in England; but as the fame caufes do not ceafe to act more or lefs every where, the effects which neceffarily follow from them will likewife become more and more fenfible, unlefs continual care be taken to prevent them. The waters of all rivers and canals are from time to time muddy: their ftreams, particularly during rains and floods, carry along with them earth and other fubftances which fubfide in thofe places where their currents are the leaft, whereby their beds are continually raifed: fo that the fucceffive increafe of inundations in rivers, and of unfitnefs for navigation in canals, when they are neglected and left to themfelves, is a natural and neceffary confequence of the ftate of things, which no intelligent perfon can be at a lofs to account for; and yet I have known whole countries remain in this habitual ftate of negligence to their very great detriment.

36. Having thus fhewn the principal accidents which rivers and canals are liable to, with the caufes of them, I

fhall proceed to point out the moft efficacious methods of
preventing them, or at leaft of diminifhing their effects.
Perhaps it would have been more proper to have deferred
doing this till I fhould have faid all I have to fay upon
the nature of rivers and canals: however, I fhall forego
the more fcientific order of things, for the fake of
bringing the means of remedying the accidents and
inconveniences which happen, nearer to the caufes
that produce them, whereby their connexion and effi-
cacity may be better judged of.   For this end, I fhall
here lay down, briefly and in general terms, the methods
moft proper for the purpofe in queftion.   They flow immediately from the principles already laid down in this
effay, and do not need many words to make them com-
pleatly underftood.

37. A work of this kind, if it is properly conducted,
muft be begun at the lower end of the river or canal;
that is to fay, at that end where their waters are difcharged into the fea, or where they fall into fome other
greater river or canal, from whence their waters are carried off without farther hindrance. If it is a river whofe
bed, by being filled up with mud, fand, or other obftacles, and by being otherwife become irregular in its
courfe, is thereby often fubject to inundations, and incapable of internal navigation, the point, from which

3   the

the work muſt be begun and directed throughout all
the reſt of the channel, is from the loweſt water-mark
of ſpring tides on the ſhore at the mouth of the river;
or even ſomething below it, if it can be done; though
this part will ſoon fill up again by the ſand, mud, &c.
which the tides ceaſe not to roll in.

If it is a canal whoſe bed is be to dug anew, or one already made, which is to be cleaned and deepened from
the ſea ſhore or ſome large river back into the country,
and where no declivity is to be loſt; as is the caſe in all
flat countries: the work muſt be begun, and the depth of
the whole channel directed, from the low water-mark of
ſpring tides, if the mouth is to the ſea, or from ſuch a
depth in the channel of the river, if the canal falls into
one, that there may be ſuch a communication of water
from the canal to the river, in all ſituations of the current, as may let boats freely paſs from one to the other.
This, of courſe, muſt alſo direct the depth of the floor of
the laſt ſluice towards the mouth of the canal, be it to the
ſea or into a river. If the bottom or floor of a ſluice already conſtructed be too low, it will ſoon fill up with ſand
or mud, and thereby hinder the gates from opening, un-
leſs it be continually cleaned out; if, on the contrary,
this floor be too high, and in a canal whoſe natural declivity is too little for the free current of the water, as is

4 H 2   generally

generally the cafe in Holland and Flanders, all depth of
the bed of the canal below the horizontal level of the
bottom of the fluice will ferve to no manner of purpofe,
either for navigation, or for carrying off the back-waters,
but will foon fill up with mud, in fpite of all means ufed
to the contrary, except that of digging it continually
anew to no manner of purpofe; as is evident from the
reafons given above (N° 33. 34.).

38. Setting off from this determinate point, at the
mouth of a river, or at the bottom of the laft fluice upon
a canal, which are to be cleaned and deepened; the work
muft be carried on, in confequence, uniformly throughout their whole courfe backwards into the country as far
as is found neceffary for the purpofes intended.   This is
to be done after the following manner :

1ft, One muft dig up and carry away all irregularities
in the bottom and fides of the bed, fuch as banks of fand
and mud, rocks, ftumps or trunks of trees, and whatever
elfe may caufe an obftacle to the regular motion of the
water, and to the free paffage of veffels upon it.

2dly, If the declivity of the bed fhould be ftill too
little to give a fufficient current to carry off the water as
often and as faft as is neceffary, the whole bed itfelf muft be
regularly deepened, and what is dug out from the bottom muft be laid upon the fides, to render it narrower in
                                    proportion

proportion to its depth. The reason of this is evident from all that has been said.

3dly, Wherever the banks are too low to contain the stream in all its situations, they must be sufficiently raised; which may be conveniently done with what is dug out from the bed: and the whole being covered with green turf will render these banks firm and solid against the corrosion of the water. It is proper at all times to lay upon the banks what is dug from the bed, by which they are continually strengthened against the force of the current.

4thly, It is often necessary to diminish the windings and sinuosities in the channel as much as possible, by making new cuts whereby its course may approach towards a right line. This is a great resource in flat countries subject to inundations; because thereby all the declivity of a great extent of the river, through its turns and windings, may be thrown into a small space by cutting a new channel in a straight line; as may generally be done without obstacle in such countries as I am speaking of, and hereby the velocity of the current will be very greatly augmented, and the back-waters carried off to a surprizing degree, as is evident from what is said above in N° 29.

5thly,

5thly, Wherever there is a confluence of rivers or canals, the angle of their junction muſt be made as acute
as poſſible, or elſe the worſt of conſequences will ariſe
from the corroſion of their reſpective ſtreams; what
they carry off from the ſides will be thrown into irregular banks in the bottom of the bed. This acute angle of
junction may always be procured by taking the direction
at ſome diſtance from the point of confluence.

6thly, Wherever the ſides or banks of a river are liable to a more particular corroſion, either from the confluence of ſtreams, or from irremediable windings and
turns in the channel, they muſt be ſecured againſt it as
much as poſſible by *weirs:* for this corroſion not only
deſtroys the banks, and alters by degrees the courſe of
the river, but alſo fills up the bed, and thereby produces
all the bad effects we have ſpoken of above in N° 33.
34. &c.

7thly, But the principal and greateſt attention in digging the beds of rivers and canals muſt be had to the
quantity and form of their declivity. This muſt be done
uniformly throughout their whole extent, or ſo much of
it as is neceſſary for the purpoſes in hand, according to
the principles laid down above (in N° 29 and 30.) Conformable thereto, the depths of their beds, and of the
floors of their ſluices, at the mouths whereby they diſ-
									charge

charge their waters, being fixed according to what we have faid in N° 37. the depth of the reft of the beds, and the quantity of declivity therein, muft be regulated in confequence thereof, fo as to increafe regularly the quantity of declivity in equal fpaces the farther we recede from their mouths, and proceed towards their fources or to the part where the regular current is to take place.

If the depth and volume of water in a river or canal is confiderable, it will fuffice, in the part next the mouth, to allow one foot perpendicular of declivity through fix, eight, or even, according to DESCHALES [d], ten thoufand feet in horizontal extent; at moft it muft not be above one in fix or feven thoufand. From hence the quantity of declivity in equal fpaces muft flowly and gradually increafe as far as the current is to be made fit for navigation; but in fuch a manner, as that at this upper end there may not be above one foot of perpendicular declivity in four thoufand feet of horizontal extent. If it be made greater than that in a regular bed containing a confiderable volume of water, the current will be fo ftrong as to be found very unfit for the purpofes of navigation, as will appear hereafter, when I come to inveftigate the quantity of declivity in feveral rivers, the degree of fwiftnefs of whofe currents is well known.

[d]. De Fontibus et Fluviis, prop. 49.

39. I

39. I dare boldly affirm, from the certain principles of hydrodynamics laid down in this effay, that if the
above mentioned things (N° 37. 38.) were carried into
execution in a proper manner; the velocity of currents
and the acceleration of motion of the waters in rivers,
and in canals when their fluices are open, might be increafed to any degree that can be required for opening
their beds, and for preventing inundations during great
rains or fudden floods: by carrying off more fwiftly the
great acceffion of water which then takes place. It
would not be difficult, by thefe means, to increafe
the velocity of the current to double and triple what
it is in rivers and canals, whofe beds for a long fpace of
time have been left to themfelves. There is not, perhaps, a country on earth but what might be freed from
inundations by thefe means. But it may be objected, that
if all I have advifed was put in execution, even in the
flatteft countries, the currents of rivers (for canals fhut up
with fluices are here out of the queftion) would become
incommodious, if not unfit, for navigation, efpecially
againft their ftreams. This objection would be of weight
if it was not evident that the various means which I have
pointed out may be executed in whole or in part, to a
certain degree, and no farther than neceffary for the purpofes required. But, as it is certain that a ftrong and
regular

regular current in a river is the best of all means for keeping it open and deep, and for preventing the formation of banks in the bed by the subsidency of mud, &c. which it does not allow time to precipitate; I leave it to be considered, whether it is better to have a free and open navigation something incommoded by the strength of the current, or to have soon no navigation at all, without repeatedly digging the bed anew.

40. I shall not here enter into the mechanical part of the methods of digging and cleaning canals, rivers, and sea ports, or into any description of the machines and instruments necessary for that purpose. The subject would lead me much too far: besides all these things may be found much at length in most of the authors who have wrote upon hydraulic-architecture, such as BARATTERI, CORNELIO MEYERI, GUGLIELMINI, and *a notorious anonymous French plagiary,* who has taken from MEYERI, without ever naming him, almost all that is contained in his book, published at Paris in 1693, and at Amsterdam in 1696, in octavo, under the title of *Traité des Moyens de rendre les Riviéres navigables.* But the author who has treated this subject with the greatest care, and most at length, is the celebrated BELIDOR, in his *Architecture Hydraulique,* 4 vol. in quarto. To these may be added a late memoir of M. FORFAIT of

Rouen, vice-architect of the French navy, which gained
the prize of the Royal Academy of Sciences and Belles
Lettres of Mantua, for having given the beſt ſolution of
a problem propoſed by that Society in 1776, in the following terms: " *To indicate the beſt and cheapeſt method*
" *of freeing navigable canals from banks of ſand and*
" *earth formed in their beds which render them too ſhal-*
" *low.*" This piece, printed at Mantua, by Pazzoni, in
1778, contains ſixty-three pages in quarto, and is divided into two parts; the firſt contains the means of
preventing the formation of banks in navigable canals;
and the ſecond offers divers methods for remedying
them when they are already formed. For this purpoſe
the author propoſes ſix different machines of his own
invention: the firſt may be employed in rivers near the
ſea, and ſubject to the ebb and flow of the tides; the ſecond may be uſed in thoſe where the waters are always
nearly of the ſame height and velocity; the third and
fourth are to be uſed in thoſe places where the violence of
the currents corrode the beds; and the two laſt ſerve to
break up the banks of ſand or earth formed in the bottom, and to carry off all heterogeneous bodies ſunk in
the river, which cauſe an obſtacle to the current. It
would be difficult to give a juſt idea of theſe machines
without the help of the ſix plates which accompany the
                                                    piece;

piece; but as this production of a foreigner has been crowned in Italy, the country of all others in which, from all antiquity, the fcience of rivers and canals has been moft cultivated, we cannot well doubt of its merit, or that it is worthy of a tranflation into our own language.

V. *Other confiderations on the nature of rivers and inundations.*

41. Rivers flowing along plains, as well as through vallies, have naturally their beds in the loweft part of the ground comprized between the oppofite hills or mountains: neverthelefs, the furface of the water of a river in the midft of a plain is often higher than the furface of the grounds adjacent to the banks of the river. This proceeds from the continual fubfiding of the mud, &c. brought down by the ftream during floods; the waters in that cafe ufually overflowing the banks fpread themfelves over the plain, where they lofe a great part of the fwiftnefs of their current, which contributes greatly to the fubfiding of the mud they contain; fo that the farther they flow upon the plain, the clearer they grow, and the lefs remains to fubfide. From hence the greateft precipitation of mud muft be in the parts of the plain neareft the fides of the river, which in length of time will raife

thefe

thefe grounds above the reft of the plain. Again, the waters in the bed itfelf depofing inceffantly a part of the mud, &c. brought down by the ftream, muft continually, though infenfibly (for a long fpace of time) raife the channel and banks of the river above the reft of the plain. Thefe caufes may at laft contribute to the forming of an intire new bed for the river; for as all rivers carry down in their ftreams more or lefs mud and other heterogeneous matters, which do not fubfide regularly in all parts alike, but muft precipitate fafteft where the current is floweft; there muft accumulate by little and little in thefe parts fuch banks of fand and mud, as will in time hinder the current of the waters, make them reflow, and at laft totally change their direction.

Canals are ftill more fubject than rivers to have their beds raifed and their currents ftopped by the fubfiding of mud and heterogeneous matter in different places, and efpecially juft above their fluices; becaufe of the fudden ftagnation of the water which firft begins there as often as the fluices are fhut: and as there is a neceffity for keeping them for the moft part fhut, the ftagnating waters in their beds muft precipitate their mud, &c. in a much greater proportion than can be done in the currents of rivers, which are in a continual motion towards the fea.

42. I

42. I call *center of the current*, or, more properly, *line of greateſt current,* that *line which paſſes through all the ſections of a river, in the point where the velocity of the current is the greateſt of all.* We have ſeen above (N° 25.) that if the current of a river is regular, and in a right line, its center or line of greateſt velocity will be preciſely in the center of the ſections (N° 6.): but, on the contrary, if the bed is irregular and full of turns and windings, the center or line of greateſt current will likewiſe be irregular, and often change its diſtance and direction with regard to the centers of the ſections through which the waters flow, approaching ſucceſſively, and more or leſs, to all parts of the bed, but always in proportion and conformably to the irregularities in the bed itſelf.

This deviation of *the line of greateſt current* from the centers of the ſections through which it paſſes, is a cauſe of many and great changes in the beds of rivers, ſuch as the following:

1ſt, In a ſtraight and regular bed, the greateſt corroſion of the current will be in the middle of the bottom of the bed; becauſe it is that part which is neareſt to the line of greateſt current, and at the ſame time which is moſt acted upon by the perpendicular compreſſion of the water. In this caſe, whatever matters are carried off from the bottom will be thrown, by the force of the current,

current, equally towards the two fides, where the velocity of the ftream is the leaft in the whole fection.

2dly, If the bed is irregular and winding, the line of greateft current will be thrown towards one fide of the river, where its greateft force will be exerted in proportion to the local caufes which turn it afide: in fhort turns of a river there will be a gyration, or turning round of the ftream, by reafon of its beating againft the outer fide of the angle; this part will be corroded away, and the bottom near it excavated to a great depth. The matters, fo carried off, will be thrown againft the oppofite bank of the river where the current is the leaft, and produce a new ground, called an *alluvion.*

3dly, Inequalities at the bottom of a river retain and diminifh the velocity of the water, and fometimes may be fo great as to make them reflow: all thefe effects contribute to the fubfiding of fand, earth, and other matters thereon, which ceafe not to augment the volume of the obftacles themfelves, and produce fhallows and banks in the channel. Thefe in time, and by a continuance of the caufes, may become iflands, and fo produce great and permanent changes and irregularities in the beds of rivers.

4thly, *The percuffions of the center of the current* againft the fide of the bed are fo much the greater as they are
made

made under a greater angle of incidence; from whence it follows, that the force of percuffion, and the quantity of corrofion and of detriment done to the banks and weirs of rivers, and to the walls of buildings made therein, and which are expofed to that percuffion, *are always in a direct compound proportion of the angle of incidence, of the greatnefs and depth of the fection together, and of the quantity of velocity of the current.*

5thly, It may happen in time, that the excavation of the bottom, and the corrofion of the fides, will have fo changed the form of the bed as to bring the force of per-cuffion into equilibrium with the velocity and direction of the current; in that cafe, all farther corrofion and ex-cavation of the bed ceafes (N° 31.)

6thly, This gives the reafon why when one river falls into another almoft in a perpendicular direction, and makes with it too great an angle of incidence, this di-rection is changed in time, by corrofions and alluvions, into an angle much more acute, till the whole comes into equilibrium.

7thly, So great and fuch continued irregularities, from local caufes, may happen in the motion of a river, as will intirely change its ancient bed, corrode through the banks, where they are expofed to the greateft violence of

percuffion

percuffion of the ftream, and open new beds in grounds lower than what the old one is become.

8thly, Hereupon the ftate of the old bed will entirely depend on the quantity of water, and on the velocity and direction of the current in the new one; for immediately after this divifion of the waters into two beds is made, the velocity of the current in the old one will be diminifhed in proportion to its lefs depth. In confequence thereof, the waters therein will precipitate more of their mud, &c. in equal fpaces than they did before; which will more and more raife up the bottom, fometimes even till it becomes equal with the furface of the ftream. In this cafe, all the water of the river will pafs into the new bed, and the old one will remain intirely dry. It is well known, that this has happened to the Rhine near Leyden, and to many other rivers.

9thly, Hence the caufe of the formation of the new branches and mouth, whereby many great rivers difcharge their waters into the fea.

43. But in proportion as *a river, that has none of thefe obftacles in its bed,* approaches towards its mouth, we fee the velocity of its current augment, at the fame time that the declivity of the bed diminifhes, the caufes of which have been explained above (N° 29.). It is for this reafon, that inundations are more frequent and confi-

I   derable

derable, and do more damage in the interior parts of a country, than towards the mouths of most rivers.

In the Po, for example, the height of the banks made to keep in the waters diminishes as the river approaches to the sea. At Ferrara they are twenty feet high; whereas nearer the sea they do not exceed ten or twelve feet, although the channel of the river is not larger in one place than in the other.

44. The mouths of rivers, by which they discharge their waters into the sea, are liable to great variations, which produce many changes in them.

1st, The velocity and direction of the current at these mouths are in a continual variation, caused by the tides, which alternately retard and accelerate the stream.

2dly, During the flowing of the tide, the current of the river is first stopped, then turned into a direction intirely contrary throughout a considerable extent; if we may believe M. DE BUFFON, there are rivers in which the effect of the tides is sensible at 150 or 200 leagues from the sea.

3dly, This state of things is a cause of a great quantity of sand, mud, &c. being precipitated and accumulated in the channel near the mouth. This continually raises and widens the bed, and at last changes it intirely into a new place, or at least opens new mouths to dis-

VOL. LXIX. 4 K charge

charge the waters at. The Rhine, the Danube, the
Wolga, the Indus, the Ganges, the Nile, the Miffifippi,
and many other rivers, are inftances of this.

4thly, All thefe effects are lefs fenfible at the mouths
of little rivers, as their currents oppofe no fenfible obftacle to the flowing of the tides; fo that the ebb carries off
again what the flow had brought in.

45. Whenever the courfe of a river throughout a con-
derable extent of country approaches towards a right
line, its current will have a very great rapidity; and the
velocity wherewith it runs diminifhing the effect of its
natural gravitation, the middle of the current will rife
up, and the furface of the river will form a convex curve
of fufficient elevation to be perceived by the eye; the
higheft point of this curve is always directly above *the
line of greateft current* in the ftream.

On the contrary, when rivers approach near enough
to their mouths for a fenfible effect to be produced in
them by the flowing of the tides; and alfo when in
other parts of their courfe they meet with obftacles at the
fides of their channel: in both thefe cafes the furface of
the water at the fides of the current is higher than in the
middle, even though the ftream be rapid. In this fituation of things, the furface of the river forms a concave
curve, the loweft point of which, or that of inflexion, is

3   directly

directly over *the line of greatest current.* The reason
thereof is, that there are in this case two different and
opposite currents in the river; that whereby the waters
flow towards the sea, and preserve their motion therein
even to a confiderable diftance; and that of the waters
which remount, either by the flowing of the tide, or by
their meeting with local obftacles, which form a *counter
current,* fo much the more fenfible as the flowing of the
tide is ftronger, or as the percuffion of the water is made
againft greater obftacles, and in a direction nearer to a
perpendicular to them. From both thefe caufes, the
greater of which by far is that of the tides, the water near
the fides of the channel, where the velocity of the defcending ftream is naturally the leaft (N° 25), takes a
contrary direction, and runs back in the river, while that
in the middle continues to flow on towards the fea.
This counter current is what the French call *a remous.*

An ifland in the middle of a river produces the fame
effect as obftacles at the fides, regard being had to the
difference of fituation of each.

Eddies and *whirlpools* in rivers, in the center of which
there appears a conical or fpiral cavity, and about which
the water turns with great rapidity and fucks in whatever
approaches it, proceed in general from the mutual percuffion of thefe two counter currents; and the vacuity in

the

the middle is produced by the action of the centrifugal
force, whereby the water endeavours to recede, in a direct
ratio of its, velocity, from the center about which it
moves.

46. If rivers perfevered always nearly in the
fame ftate (N° 5.) the beft means of diminifhing the
velocity of the current, when it is found too great for
the purpofes of navigation, would be by widening the canal: but as all rivers are fubject to frequent increafe
and diminution, and confequently to very different degrees of velocity and force in the current, this method
is liable to produce very detrimental effects; for, when
the waters are low, if the channel is very large in proportion, the ftream will excavate a particular bed, which,
according to the irregularities of the bottom, will form
various turnings and windings with regard to the principal bed; and when the waters come to increafe, they will
follow, to a certain degree, the directions which the bottom waters take in this particular bed, and thereby will
ftrike againft the fides of the channel, fo as to deftroy the
banks and caufe great damages.

It would be poffible to prevent in part the bad effects
proceeding from the current ftriking againft the banks,
by opening, at thofe places where it ftrikes, little gulfs
into the land, dug in fuch a form and direction as that the
ftriking

ftriking current fhould enter and circulate therein, fo as to deftroy, or at leaft greatly diminifh, its velocity. This effect would be felt for a confiderable way down the river.

This fame method might probably be ufed with fuccefs againft the deftruction of bridges, weirs, &c. by the violence of the ftream during floods. Such gulfs being dug into the outer fide of thofe turnings in the river which are immediately above the place to be fecured from the violence of the ftream, would fucceffively diminifh its velocity, its force and dangerous effects, a confiderable way down the river. It is true, this method might contribute to produce an overflowing of the river upon the grounds adjacent to thofe artificial gulfs, this being a natural confequence of the decreafe of the velocity of the current in thofe places; and it would remain to be confidered whether thofe local inundations, or the danger of deftruction of the bridges or edifices in the river, were the leffer evil.

47. The nature of inundations, and the manner of their formation, merit a particular attention in this place.

While the volume of water in the *bed* of a river increafes, the velocity of the current increafes in proportion, as has been repeatedly fhewn above (N° 13. 18. 20.

23. 27. 28. 29.). But from the moment that part of
this water overflows the bed, the velocity thereof begins
to diminifh (N° 41.) and does fo more and more, the
farther it flows and fpreads on the plain. So that the
overflowing being once begun, it is a natural confequence, that the inundation fhould continue for feveral
days; for though the volume of water brought down by
the flood during that time fhould decreafe, yet, as the
quantity of what runs off decreafes likewife, from the
great decreafe of velocity in what overflows the plains,
it will continue to produce the fame effect as if the volume of water coming down had not diminifhed, until
the whole of the ftream be every where contained again
within the bed of the river. When that is become the cafe,
the waters that have overflowed the plain will decreafe
thereon, by gradually and flowly running off, and alfo by
evaporation, till they wholly difappear. If this was not
fo, we fhould fee rivers overflow for an hour or two,
and then return again within their beds, a thing contrary to general obfervation; for we conftantly fee inundations, once begun in flat countries, laft for feveral days
together, although in the mean while the rain ceafes,
and the quantity of water coming down diminifhes. This
muft be the cafe, becaufe as the overflowing diminifhes
the velocity, and confequently the quantity of water
carried

carried off, it has the fame effect as if a greater quantity ſtill continued to come down.

It may not be uſeleſs to remark here, that what we have often ſaid in this eſſay becomes evident from theſe obſervations on nature, as well as from the principles laid down in it; to wit, that the moſt direct and efficacious method of preventing inundations is by deepening the bed and raiſing the banks of the river.

It may likewiſe be obſerved, with regard to inundations, that if the wind blows directly contrary to the current of the river, the overflowing will be greater than it would have been otherwiſe, becauſe this accident diminiſhes the velocity of the ſtream: but, on the contrary, if the winds blow in the ſame direction with the current of the river, the inundation will be leſs than otherwiſe, and ſooner at an end; becauſe this accidental cauſe augments the velocity of the ſtream.

VI. *On the confluence of rivers, and on the ſeparation of the ſame river into divers branches and mouths, with the effects thereof upon the velocity of currents, inundations, &c.*

48. All great and long rivers receive into their beds many others of different magnitude throughout the extent

tent of their courfe. This is evident to every one who
only cafts his eyes over a map. The Rhine and the Po,
in particular, receive each above one hundred others
great and fmall; the Danube above two hundred; the
Wolga as many; the river of Amazons receives into its vaft
bed a prodigious number, fome of which are five or fix
hundred leagues in length, and are of fuch a depth and
breadth as would make them elfewhere pafs for capital
rivers. M. DE BUFFON [e] gives a lift of the more confiderable of thofe which fall into other great rivers
throughout the known world. Many curious particulars
may be feen in VARENIUS's General Geography, part I.
chap. XVI. concerning rivers; but of a nature which
does not enter into my plan. The works themfelves are
in every body's hands, and may be confulted by thofe
who pleafe.

This confluence of rivers is fo neceffary for propagat-
ing the motion of the water throughout a long courfe,
and for renewing and accelerating from time to time its
velocity, which otherwife would be too greatly diminifhed by the refiftance of fo many obftacles as they
meet with in their way, that, as we have faid above (N°
28.) after Signor GUGLIELMINI, it can only be attributed

[e] Hift. Nat. tom. II. p. 75, 76.

ta

to the infinite wifdom of the Author of Nature, in the original difpofition of things.

49. We have feen above (N° 18. 27. 28.) that the increafe of a river or canal by the new waters which it receives, *is univerfally in an inverfe ratio of the new velocity which is acquired therefrom.* If this velocity is greater, the increafe of the fection of the new ftream will be lefs in proportion, and *vice verfâ.* It follows from hence, that it is poffible for one river or open canal to fall into another river or open canal of equal magnitude with itfelf, and yet the fection of the current in the common bed after their confluence fhall be no greater than it was in each of them before their junction. It is certain that this will be the cafe as often as *the confluence of the two augments the velocity of the common current in the fame proportion with the increafe of the volume of waters;* both the greater rapidity of the current, and the greater volume of water in the bed after the junction, ferving to deepen it in proportion to its breadth, will contribute towards the above effect. Another caufe will likewife add thereto; to wit, that inftead of the refiftance from the attraction, friction, and other obftacles, in two beds, which give two bottoms and four fides, there are, after the confluence, only thofe of one bed, of one bottom and two fides. Moreover, the center of the fection in the common bed

is farther from the bottom and fides thereof, than it is in the feparate beds. All thefe caufes, in proportion to their refpective quantities, contribute to accelerate the velocity of the common ftream.

50. It is not lefs certain, that in rivers which bring down a great abundance of water, the more the velocity and difcharge thereof at their mouths are retarded and diminifhed by the tides, the winds, the rolling in of the fea, &c. the more will the back-waters increafe in height, and endanger overflowing the inner parts of the country. This is evident, becaufe the decreafe of velocity in the current, and the increafe of height of all the back-waters that are affected thereby, are in a reciprocal inverfe ratio one of another (fee above N° 32.)

Nature itfelf teaches us a method of preventing, or at leaft of diminifhing, this effect. We fee all great rivers overcharged with a vaft volume of water divide, when they come near the fea, into different branches and mouths, whereby the fuper-abundance of their waters is difcharged. This is the cafe with the Scheld, the Rhine, the Rhone, the Po, the Danube, the Wolga, the Euphrates, the Indus, the Ganges, the Nile, the Niger, the Oroonoko, the River of Amazons, and with almoft all other great rivers.

7

This

This feparation and difperfion of the too great quantity of water into feveral channels is one caufe of their feldom overflowing the country near their mouths *(f)*, becaufe it gives a greater depreffion and declivity to the furface of the current, and thereby facilitates the running down of the waters from the interior parts of the country, forafmuch as their beds are every where regular and free from obftacles to their current.

51. Notwithftanding the apparent oppofition to what has been faid in feveral other parts of this treatife, I repeat again, that this divifion and difperfion of the waters into feveral branches and channels *when there is fuch an abundance of it as is fufficient to keep up the velocity both in the old and new channels,* augments the declivity, and thereby facilitates the running off of all the back-waters from the inner parts of the country, as far as the bed is regular and free from obftacles, according to what is laid down above (N° 38.).

But whenever this fuper-abundance of waters, fufficient for keeping up the velocity in each channel nearly to what it was before the feparation or divarication, fhall be found wanting, it is certain, that this divifion and difperfion of the waters into feveral channels will only ferve to diminifh the velocity of the current in each, whereby as much or more difcharge of the water, and confequently of declivity for the running off of the back-waters, may be loft, as has been gained by the feparation into different beds.

This difadvantage may be eafily remedied in thofe new channels and mouths of rivers which are dug by hands, and have fluices placed in them at the point of feparation from the original bed; for thefe fluices of communication need be opened only when there is a fuper-abundance of water in the river, fufficient to keep up the velocity in each of the channels; at other times they may be kept fhut, and the waters retained in their original bed.

52. It was for this purpofe, of preventing the damages proceeding from immoderate inundations, that the ancient Egyptians dug vaft lakes, and made fo many canals and fluices of communication between the Nile and thofe lakes, and from thence to the fea; that they might thereby be able to difcharge the waters into thofe refervoirs if they came down in too great abundance, or let them off again from thence upon the land, if the quantity of the natural inundation at any time was lefs than what was neceffary for the good of the country. By thefe means ancient Egypt was always mafter of its waters.

I   It

It is well known that it rains feldom in that country, and that the Nile by its regular inundations waters the land, by bringing down upon it the rains and melted fnow from the high mountains of Abyffinia. HERODOTUS [g] and DIODORUS SICULUS [h] have left us defcriptions of the immenfe labours of the inhabitants to govern and multiply fo beneficent a river, the particulars whereof are too well known to be repeated in this place. By thefe means Egypt became the granary of the world for above two thoufand years, and reimburfed, with immenfe advantage, the firft expences.

RICCIOLI [i] affures us, that the ancient Perfians did the fame thing with regard to the *Euphrates,* and for the fame end. He adds, moreover, *Sic ubi Cyrus Gangem in Alveos 460 difperfit, minora damna ex Gangis alluvionibus campi perpeffi funt;* but I am totally at a lofs to find upon what authority he grounds this laft affertion, for I never read that any CYRUS penetrated as far into India as the mouth of the *Ganges,* much lefs reigned fo long over that country as to perform the vaft work which RICCIOLI fpeaks of.

(g) In lib. II.
(h) Biblioth. I. II. c. I.
(i) Geogr. et Hydrogr. I. VI. cap. xxix. p. 248, 249.

PLINY

PLINY [k] fays, with regard to the different mouths of
the Po, *Omnia ea flumina foffafque primi à Sagi fecere
Thufci, egefto amnis impetu per tranfverfum in Atriano-
rum Paludes, quæ Septem Maria appellantur.* Thefe
feven lakes difcharged their waters into the fea by
feven mouths, which PLINY names in the fame place.
All this was apparently done that the river might do
lefs damage to the adjacent countries by its frequent inundations.   PLINY adds, *His fe Padus mifcet, ac per hæc
effunditur, plerifque, ut in Ægypto Nilus, quod vocant
Delta.*

To thefe examples, drawn from ancient hiftory, might
be added many modern ones, if the things in queftion
had need of further proofs.   Thus, both nature and the
experience of a long feries of ages teach us, that the feparation of a river into feveral beds, by new branches
and mouths, is a means of diminifhing inundations in
the inner part of the country; but that this takes place
only when there is a fufficient abundance of water in the
river to fill the new beds and channels fo far as to prevent
the velocity of the currents therein from being notably
diminifhed from what they were before the divifion.

[k] Hift. Nat. l. III. cap. xvi.

SEC

SECTION II.

Laws of the meeting of opposite currents, with the application of them to fluices.

53. When two equal currents of homogeneous fluids meet in oppofite directions, there is firft a fwelling and rifing up of them at the point of rencounter; then follows a revulfion and counter current of each equally back again, fo as to bring the whole to an equilibrium.

54. If the two oppofite currents are unequal, either in force or in quantity, or in both, there will ftill be a fwelling and revulfion of each back again, but it will be diminifhed in the greater current, and augmented in the leffer, by the quantity by the which the one furpaffes the other; and the point of rencounter of the two currents will have a flow and progreffive motion in the direction of the ftronger, the degree of velocity thereof being always in a direct ratio of the force and quantity of the one above the other,

55. If the fluids in oppofite currents be not homogeneous, as is the cafe between fea and river water, that which has the leaft fpecific gravity will fwim upon the other, and continue to follow its firft direction, until

fuch

fuch time as the heavier fluid fhall have communicated its motion to all the parts of the lighter. But the lighter fluid will not lofe its former motion and direction at once, but *in a decreafing feries, the law whereof will vary according to the greater or lefs difference of fpecific gravity in the two fluids,* until the whole of the lighter has acquired the velocity and direction of the heavier which buoys it up.

The *time* and *fpace* required for a greater current of falt water to communicate its motion and direction to an oppofite one of frefh water will be but very little, fince they differ in fpecific gravity only $\frac{3}{73}$ parts that the falt is heavier than the frefh. It would require much greater between water and oil, and ftill much more between quickfilver and oil, and fo on. The elements for determining them in every cafe might be found by a proper number of experiments.

56. Let the two currents be equal or unequal in force and velocity *but nearly of the fame fpecific gravity,* if we fhould fuppofe at the fame time that their furfaces are not upon a level, but that the one is higher than the other (as is conftantly the cafe in all fluices that open to the fea, except at the moment when the furface of the tide is upon a level with the furface of the water in the canal behind the fluice); this circumftance entirely changes both the cafe and the effects. It is certain, on

this

this fuppofition, that the overplus of velocity and elevation in the higher current, though it fhould be the leffer,
will make the waters in the lower and greater current
reflow upon themfelves until they come to a level and
equilibrium with thofe in the upper current; fince
thefe, by the laws of univerfal gravitation, cannot flow.
back from a lower to a higher level, but muft defcend
according to the declivity of the furfaces. If the currents
are of *very different fpecific gravities,* they will come to
an equilibrium according to the law laid down above (N°
55); but their greater or leffer quantity and velocity
will produce little or no effect in this cafe.

57. Now as the running of two currents in oppofite
directions, after their rencounter, and beyond the limits
laid down above (N° 55.), is incompatible with, and con-
tradictory to, the laws of nature, and confequently impoffible; we may draw this ufeful conclufion, which becomes important during inundations, and efpecially
during the annual overflowing of the low grounds in
flat countries; to wit, that if the fluices next the fea
againft which the tide flows be fhut only a quarter of
an hour before the *flood* has rifen to the level of the
water in the canal, not a drop of falt water can enter
the faid canal, nor even into the fluice itfelf; becaufe
both the progreffive motion of *the point of rencounter of*

the two currents, and the *over-fwimming* of the frefh
water upon the falt, will be always without the fluice and
towards the fea, fo long as the furface of the tide is below
the level of the water in the canal. Many fluice mafters,
for want of knowing or confidering this, are accuftomed
to fhut their gates next the fea a little after *half flood,* under the pretence of preventing by this means the falt water from getting into the canal, and communicating
thereby with the waters that overflow the low grounds
in many places during winter, which would be of great
detriment to the foil. Through this falfe perfuafion,
they lofe no inconfiderable part of that time every day,
which they might fafely employ in drawing off the waters which overflow and incommode low and flat countries almoft every winter and rainy feafon, as is the cafe
in the Dutch and Auftrian Netherlands.

SECTION IV.

*Experiments to determine the different velocities, in different
depths of water, of the fame floating body moved uniformly by an equal force.*

58. It is well known already, that for facilitating or
retarding the motion of boats, &c. in canals, the different

depths of the water, above that fimply neceffary to keep them afloat, is a thing not at all indifferent. Dr. FRANK-LIN has already treated this fubject, though perhaps not with fufficient accuracy, in a letter to Sir JOHN PRINGLE, written in the year 1769. He proves, however, that it is univerfally known among people accuftomed to work boats on canals, that there is a confiderable difference in the fwiftnefs of their motion according to the greater or lefs depth of the water therein; and that the water being low is of itfelf fufficient to retard the motion of a boat, without the keel thereof rubbing againft the bottom of the canal. The reafon he affigns for it is evident; for a boat cannot advance its own length in a canal without difplacing a quantity of water equal in mafs to the fpace which the boat occupies under the furface of the fluid. The water fo difplaced muft retrograde, and pafs under, and to the right and left, of the boat: fo that the lefs depth and breadth of water there is in the channel, the more in proportion it muft rife up and weigh againft the boat, and the more difficulty it muft find in paffing under and along fide of it, and neceffarily muft retard fo much the more the motion thereof. The refult of Dr. FRANKLIN's experiments on this fubject may be feen in the letter above mentioned.

59. Mr. NEEDHAM, Director of the Imperial Academy of Sciences at Bruxelles, being of opinion that Dr. FRANKLIN's experiments were made upon too fmall a fcale to draw any very exact inferences from them, defired me, at the beginning of the year 1775, to make a new fet of experiments upon a much larger fcale and with all poffible exactnefs; I did accordingly, and fhall here give a fhort defcription of them.

I got made, by the fhip carpenters of Nieuport upon the Coaft of Flanders, an exact model of a bilander, anfwerable in all its proportions to thofe ufed in the Low Countries. Its length was thirty-nine Englifh inches, its breadth nine inches and a half; and its depth nine inches. Its form both within and without exactly reprefented that of a bilander. At each end of it was faftened perpendicularly a round and polifhed rod, ten inches and and a half in height above the fides of the boat.

I got made likewife a wooden canal, twenty feet in length, thirty-feven inches in breadth, and fixteen inches in depth; a fection of which is reprefented in the following figure:

This

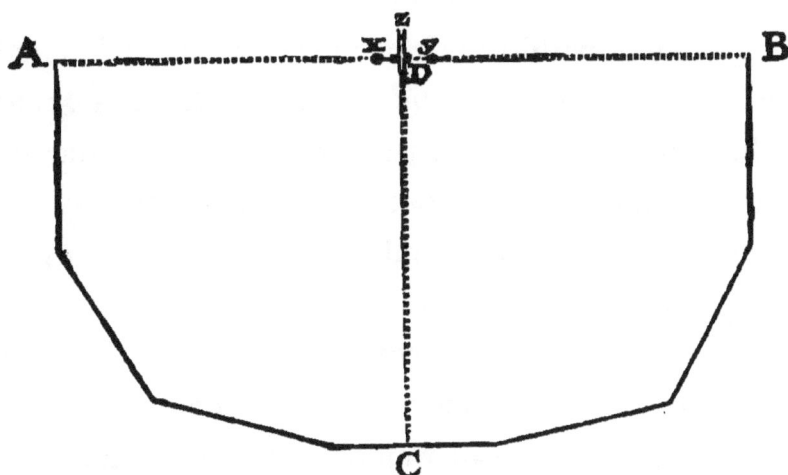

This form is that of the excavation of the canals in the Low Countries, and approaches to that of the natural beds of rivers inasmuch as they are regular. Here AB = 37 English inches, DC = 16 inches; and the length of the whole canal, as we said before, was twenty feet. z represents the section of a pulley fixed at one end of the canal, upon which passed a small cord, one end of which was tied to the round rod at the fore part of the boat, and at the other end was a piece of lead which weighed eight ounces. This served for an equable force to give an uniform motion to the boat throughout all the experiments. *x* and *y* are sections of two other cords stretched parallel to each other at about one inch and a half distance, and reaching from one end of the canal to

2   the

the other. The two round rods fixed at the ends of the
boat, moving within thefe parallel cords, ferved to make
the boat move in a right line in the middle of the canal,
without running againſt either fide, which it would have
done without this precaution. The canal itfelf was upon
an exact level, and one end of it, where the pulley was
fixed, reſted upon the fide of a well twenty-three feet
deep, twenty of which were above the furface of the
water; which gave fufficient fpace for the free and uniform defcent of the lead-weight and cord running over
the pulley, as they drew the boat from one end of the
canal to the other.

Latts, exactly divided into inches, were nailed againſt
each end of the canal within, to mark the different
depths of the water in it according as it fhould be augmented or diminifhed. The outfides of the little boat,
from its keel upwards, were likewife divided into inches.
In the infide of the boat was a quantity of fand fufficient
to fink it to fix inches deep in the water. The common
loaded bilanders in the Low Countries ufually draw fix
feet of water.

Thus the form of the wooden canal, together with its
breadth and depth, and the form and dimenfions of the
little boat, together with the depth of water it drew by
means of its ballaſt of fand, exactly correfponded with
                                                    thofe

thofe in the real canals and bilanders in the Low Countries, an inch in the one anfwering to a foot in the
other.

Clofe to the canal, and out of the way of all wind, was
fufpended a pendulum of fine waxed thread, to prevent
the variations of the atmofphere from altering its length,
which from the point of fufpenfion to the center of gravity in the lead was $39\frac{1}{5}$ Englifh inches, fo that its ifo-
chronic vibrations were exactly feconds of time.

60. It was neceffary, in order to render the experiments exact, that they fhould be made at a time when
the air was perfectly calm; for the leaft breath of wind,
during the motion of the boat, caufed great variations
and irregularities in them, which it was abfolutely neceffary to prevent, in order to be able to deduce any exact refults from them. On the contrary, in a perfect calm, the
times of the paffage of the boat, from one end of the canal to the other, were exceedingly regular, as may be feen
from the table of experiments which I give below.

By means of the pendulum I was able to meafure
the times of paffage of the boat along the canal, in all
the different depths of water, to a third or even to one
quarter of a fecond. The boat being held faft againft
the back end of the canal by the hand of an affiftant, and
then let go, it was eafy for me to perceive the precife

inftant

inftant of the beginning of its motion, to let go the pendulum at the fame moment, and to count its vibrations
till the inftant that the boat ftruck with an accelerated
force againft the fore end of the canal. As to the weight
of eight ounces fufpended at the end of the cord, and
which ferved as *a moving force* to draw the boat along
the canal, it was juft as much as fufficed to counterbalance the cord, and to put the boat in motion; lefs
weight than that would do neither: therefore I was
obliged to ufe fo much, notwithftanding the confiderably
accelerated motion it gave to the boat.

This is the whole mechanifm of the inftruments I
ufed for the experiments in queftion, and fuch were the
precautions I judged it neceffary to ufe for making them
with fcrupulous exactnefs.

In the following table, which confifts of twelve co-
lumns, the firft of them contains the different depths of
water at which the experiments were made; the ten following ones contain ten different experiments made at each
depth of water in the canal; and the twelfth or laft column is the reduction of the ten others to a mean proportional or mean refult of the whole, which is in *feconds*
of time.

Table

Table of experiments made to aſcertain the times of paſſage of the boat along the canal, or its different degrees of velocity, in different depths of water.

Depths of Water.	1ſt Exp.	2d ditto	3d ditto	4th ditto	5th litto	6th ditto	7th ditto	8th ditto	9th ditto	10th ditto	Mean Reſults.
Inches 15	″ 14½	″ 14¾	″ 14⅓	″ 14½	″ 14½	″ 14⅓	″ 14⅓	″ 15	″ 14½	″ 14⅓	″ 14⅓
14	14¾	15	15	15½	15	15	14½	14½	15	15¼	15
13	15½	15½	15½	15¾	15½	15¼	15¼	15½	15¼	15¼	15½
12	16½	16	16	15½	16	15¾	16	16	16	16	16
11	16¾	17	17	16½	16¾	17	16¾	16½	16¾	16¾	16¾
10	17	18 wind	17	17½	17	17⅓	17½	17	17	17¾	17½
9	18½	18½	18	19	19	18	18	18¾	18	18⅓	18½
8	20	20	20½	20	20	20	20	19½	20	20	20
7	23½	23	22½	23	23	23	23½	23¼	23	23	23
6½	In the experiments made with this depth of water, the boat often touched the bottom.										30 by ſuppoſition

It may be obſerved with regard to the laſt column of the above table, which contains the mean reſults or mean quantities of time which the boat takes to paſs from one end of the canal to the other in different depths of water, that it is given for the ſake of deſtroying thoſe little differences which are inevitable in practice; and it

 ſhews,

fhews, as nearly as poffible, what the true time of paffage ought regularly to be when nothing happens to dif-
turb it.

It is alfo highly worthy of remark, that the *mean
refults* contained in this laft column form a *feries of numbers regularly increafing as the depths of water, wherein
the refpective experiments were made, regularly decreafe*;
fo that *the different velocities of the floating body are in an
inverfe ratio of the refpective depths of the water in which
it floats with an equal impulfive force, and that according
to the law of the above feries.* This, perhaps, may fur-
nifh elements to calculate, pretty near the truth, the different velocities of veffels upon canals and rivers with
different depths of water in all other cafes whatfoever.
As to the conclufions to be drawn therefrom in practice,
and in the common ufes of life, they are too obvious to
need mentioning here.

SECTION V.

On the quantitity of declivity in rivers.

62. Abftracting from all refiftance and friction, fluids,
fuch as water, defcend upon planes let them be never fo
little inclined towards the center of the earth: and the
5   velocity

velocity of defcent increafes in a compound ratio of the increafe of the mafs of water, and of the greater declivity of the plane which ferves for its bed (N° 13.).

63. Water, though unaffected by any compreffion or impulfion from above, cannot remain immoveable in any bed whatever except that which makes a curve perfectly concentrical with the *terreftrial curve*; but in this, being every where equally affected by the force of gravitation, it will remain without motion any way.

64. It follows from hence, that the fprings and fources of all rivers muft be at a greater diftance from the center of the earth than one femi-diameter thereof, which is terminated at the furface of the fea; without which the waters could not run to the river's mouth.

65. Therefore, the abfolute elevation of the furface of rivers is continually diminifhed as they recede from their fprings, becaufe of the neceffary declivity of the beds of rivers towards the center of the earth; for without fome degree of this declivity the waters could not run at all, as has been faid above (N° 62. 63.).

66. The declivity of the beds of rivers cannot be a right line making a rectilinear angle with that horizontal plane which, being continued, would interfect their refpective fources; but, if it is regular, it muft be a curve which differs very little from that of the earth's furface,

and this, if the direction is in the parallels of latitude
due Eaft and Weft, is *fpherical*; but in all other directions it is a portion of an *oblate ellipfis*, on account of the
earth's being a fpheroid compreffed by its axis. Now the
horizontal plane which continued paffes through the
fprings of rivers, is always a tangent to the curves of
their beds at the point of inflexion, infomuch as thefe
are regular.

67. The quantity of abfolute declivity from the fpring
in any determinate part of a river, is that perpendicular
line drawn from the point of greateft current in that
place till it meets *the curve concentrical to the earth's
furface* which paffes through the river's fource. The declivity of the bed below the fpring is had by taking the
fame perpendicular from the bottom of the bed; as that
of the river's furface is by taking the perpendicular from
thence.

68. If a plane be extended horizontally every way
from the point of tangency to the earth's furface, or from
the point where it is perpendicular to any radius of the
earth, water will run from every other part of the plane
towards that faid point which is nearer to the center of
the earth than any other point in the whole plane.

69. The depreffion of *the curve of a river's bed*, below
*the concentrical-terreftrial-curve which interfects its
fource,*

source, being only 250 fathom perpendicular in a courfe. of 500 leagues, it will be fufficient to give a notable current in a regular bed throughout all that extent of river, as appears from what we have faid above (N° 38.). But the depreffion of *the curve of the river's bed*, below *the horizontal plane which is a tangent to its fource*, in this fame extent of courfe, is not lefs than ninety leagues perpendicular, being always *the fecant of the arc of the river's extent minus a radius of the earth in that point*.

70. It follows evidently from the above princples (N° 62—69), that the declivity and velocity of a river are lefs in proportion as the bed approaches nearer to being concentrical with the curve of the earth's furface.

71. I fhall now apply the principles laid down to determine, as near as poffible, the real quantity of declivity in different rivers, making ufe of what is already known from experiments and actual menfuration to determine the fame in all others by the comparifon of the different degrees of velocity in their refpective currents.

72. It is the general opinion of moft of thofe who have examined this fubject [k], that rivers and canals which have lefs than one foot of declivity in 10,000 feet of courfe, will have very little current, unlefs it be by means of the great abundance of their upper waters

(k) Vide DESCHALES, de Font. et Fluv.

4   which

which give motion to thofe before them by their weight
and impulfion. Without this the refiftance proceeding
from the bottom and fides of the bed, and from other
accidental obftacles (N° 21.) would equal, if not furpafs.
the ordinary caufes of acceleration (N° 20.) fo as to diminifh continually the motion of the waters, and at laft
render them almoft ftagnant (N° 22.). But nature has
prepared remedies againft this, as we have feen above
(N° 28. 29. 48.). What RICCIOLI [1] fays of the *Po*, in
that part of its courfe next its mouth, is perfectly conformable to this theory: " Sic Padus, qui a Pago *Oftellata*
" vocato, ufque ad *Adriaticum*, intervallo milliarium cir-
" citer 70, non habet libramentum majus 13 aut 14
" pedum, ita ut fingulis milliaribus ne 3 quidem unciæ
" declivitatis obveniant; unde *Paduſæ*, potius inftar ftag-
" nantibus aquis, incertiffimus effet ad defluxum curfus:
" impetu tamen impreffo à 30 et amplius fluminibus
" aut torrentibus fe in illum exonerantibus, etiamque a
" nativæ pondere aquæ ex fuperioribus et altioribus
" prope Alpes alveis decurrentis, velocitatem maximam
" acquirit."

73 From many obfervations and trials which I made
for this purpofe in the years 1773 and 1774 upon the
river Iprelee in Flanders, which comes down from the

(1) Geogr. et Hydrogr. l. VI. c. xix. p. 215. edit. Bonon. 1661.

city

city of Ipres and falls into the fea at Nieuport, having *a very moderate current* when the fluices upon it are open, I found its *mean declivity* to be nearly three fathoms four feet and eight inches in 20,000 fathoms of extent of its courfe, or very nearly one foot in a meafured Englifh mile. I fay its *mean declivity*, becaufe from what has been faid above (N° 13. 27. 28. 29.) it is plain, that a greater or lefs quantity than ordinary of water in it will add to, or take from, fomething thereof; but the declivity in each part of its bed is nearly uniform.

As the fources of this river, and thofe of the Ifere which joins it at Fort Knock, four leagues from Nieuport, are in the higher grounds of Flanders towards Houthem, Mount Kemele, Swaertfberg, Catfberg, and the other hills as far as Mount Caffel; and as the reft of their courfe is in a flat country with a very fmall defcent towards the fea, the declivity thereof may be taken as *a mean* between that of the other rivers and canals of Flanders: the *artificial canals* will have lefs, not above a fix or feven thoufandth part of their extent, or one twelfth of an inch in each eight fathoms: the rivers Lys and Efcaut, before they fall into the flat country, fomething more, after which they may have about the fame, or even fomething lefs between Ghent and Antwerp.

This

This quantity of three fathoms four feet and eight inches of declivity in 20,000 fathoms of extent, gives the proportion of the declivity to the extent as 1 to 5292, which is one line or twelfth part of an inch in $6\frac{1}{8}$ fathoms, and two feet feven inches in one French league of 2283 fathoms. Now the meafured Englifh mile containing 5280 feet, this proportion of $\frac{1}{5292}$ approaches fo very near to one foot of declivity in every meafured mile of extent, that I fhall every where reduce what I call the *mean declivity* to that quantity, as a ftandard wherewith to compare the reft.

74. In canals, all whofe fluices and vents have been kept fhut a fufficient time to render the water ftagnant throughout their whole length, there cannot be allowed above an inch or two of declivity for each mile in length, on account of the water that unavoidably runs off through the chinks of the doors of fluices, drains, &c.

75. According to the obfervations of the Abbe CHAPPE D'AUTEROCHE [m], the floor of the Hall of the Royal Obfervatory at Paris is forty-five fathoms three feet and five inches French above the level of the fea at the mouth of the Seine. According to the Abbé NOLLET, this fame floor is forty-fix fathom above the level of the Ocean, and only forty-five fathoms above the level of the Mediter-

---

(m) See Relation de fon Voyage en Siberie, tom. II. p. 406, 407. 444.

ranean fea. Again, according to the above Abbé CHAPPE, the faid floor of the Obfervatory is elevated twenty-four fathoms one foot and ten inches above the level of the river Scine at Paris; therefore the level of the Seine under the Pont Royal at Paris is twenty-one fathoms one foot and feven inches above the level of the Ocean; and fuch alfo is the quantity which Meff. CASSINI have given, from their own obfervations and experiments, for the mean height of the Scine at Paris above the level of the fea.

Now the courfe of the Seine from Paris to its mouth at Havre de Grace, by following all its turns and windings, is about 90,000 fathoms in length; therefore $\frac{90,000}{21.1.7} = 4232\frac{1}{2}$ fathoms of extent for one fathom of declivity in the river Seine, or one line in $4\frac{2}{3}$ fathoms, and confequently the proportion of its declivity to its length is as one to $4232\frac{1}{2}$. It is to be obferved, that the bed of the Seine is deep, and its current confiderably ftrong.

76. By fimilar obfervations and actual levellings made upon the river Loire by M. M. PICARD and PITOT [n], the declivity thereof in proportion to its length is found to be as one to 3174, which is one line in $3\frac{2}{3}$ fathoms. Notwithftanding this great declivity of the bed of the Loire it is obferved, that the velocity of the water therein,

(n) See Memoires de l'Açad. Royale des Sciences de Paris, pour 1730.

compared

compared to that in the Seine, is lefs than it fhould be in proportion to their refpective declivities, which is very juftly attributed to the much greater depth in proportion to the breadth of the Seine, above what is found in the Loire. This laft river is remarkably broad, and fo fhallow that in many places it is hardly navigable for boats. Now this, according to the principles laid down above (N° 27, &c.) muft very much diminifh the fwiftnefs of the current, which it fhould otherwife have from the great declivity of its bed. In confirmation of this it is moreover obferved, that in great falls of rain, which equally increafe the volume of water in both thefe rivers, the velocity in the Loire augments in a much greater proportion than it does in the Seine; and this obfervation is likewife conformable to the principles above laid down (N° 12. 28. &c.).

77. The river Doux, after paffing by Befançon, falls into the Saone above Chalon; the Saone joins the Rhône at Lyons. This river, from Befançon to its mouth in the Mediterranean fea, is one of the moft rapid in the known world: the velocity of its current is at leaft double to that of the Seine or Loire, and its courfe is almoft in a ftraight line. The difference of elevation of this river at Befançon, above that of its mouth in the Mediterranean fea, after a courfe of about eighty-fix French leagues, has

been

been found, by a long feries of barometrical obfervations, to be about feventy-five fathom [o], which gives the proportion of the declivity to the extent as one to 2620, or about one third of a line to each fathom. This is double the *mean declivity* of the rivers in Flanders: but the velocity of the current in the Rhone is at leaft triple that in the others (N° 29.).

78. From the above *data,* got from obfervations and actual menfuration, and from many others of the fame nature too long to mention here, we may deduce the following table of comparative proportions between the declivities and velocities in different kinds of rivers.

(q) See Cours de Phyfique de Para, tom. II. N° 740.

Rates

Rates or claffes of rivers and flowing waters.	Comparative degrees of the mean velocities of currents	Seconds of time wherein currents run 20 fathom.	Fathom run by the current in 1 minute of time.	Ratios of declivity compared with horizontal length.	Fathoms of length for each $\frac{1}{14}$ inch of declivity.	Diftinctive attributes of the various kinds of rivers and flowing waters.
1	0	0	0	$\frac{1}{1000}$	14	Channels wherein the refiftance from the bed, and other obftacles, equal the quantity of current acquired from the declivity ; fo that the waters would ftagnate therein, were it not for the compreffion and impulfion of the upper and back-waters.
2	$\frac{2}{3}$	180	$6\frac{2}{3}$	$\frac{1}{7000}$	8	Artificial canals in the Dutch and Auftrian Netherlands.
3	1	120	10	$\frac{1}{3200}$	6	Rivers in low and flat countries, full of turns and windings, and of a very flow current, fubject to frequent and lafting inundations.
4	$1\frac{1}{2}$	80	15	$\frac{1}{4000}$	$4\frac{2}{3}$	Rivers in moft countries that are a mean between flat and hilly, which have a good current, but are fubject to overflow : Alfo, the upper parts of rivers in flat countries.
5	$2\frac{1}{6}$	55	$21\frac{2}{3}$	$\frac{1}{3100}$	$3\frac{2}{3}$	Rivers in hilly countries, with a ftrong current, and feldom fubject to inundations : Alfo, all rivers near their fources have this declivity and velocity, and often much more.
6	3	40	30	$\frac{1}{1800}$	3	Rivers in mountainous countries, having a rapid current and ftraight courfe, and very rarely overflowing.
7	5	24	50	$\frac{1}{1000}$	$2\frac{1}{3}$	Rivers in their defcent from among mountains down into the plains below, in which places they run torrent-wife.
8	8	15	80	$\frac{1}{1700}$	2	Abfolute torrents among mountains.

I fhould

I fhould think it quite fuperfluous to give any explanation of a table fo clear and intelligible as the above; and fhall only remark upon it that *the comparative degrees of the mean velocities of the refpective currents* in the fecond column are the refult of obfervations and experiments, the method of making which has been given above (N°26.): but as the velocity of rivers is very different in different feafons of the year, which augment or diminifh greatly the mafs of waters in their beds, *a mean* has been kept to, as much as poffible, in the above table.

By taking the degree of velocity of the current in any river, a thing fo eafy to be done; and obferving its other characterifics as laid down above under the title of *diftinctive attributes*, it will be eafy to judge very nearly of *the quantity of declivity in that part of the river.*

79. After carefully comparing what has been faid in the relations of travellers, and in the beft treatifes of geography, upon the principal rivers in the known world, I fhould be inclined to clafs them in the following manner, particularly entreating at the fame time that my opinion about it may be regarded as fimple conjecture, which I leave to be rectified by thofe better acquainted with the matter than it is poffible for me to be.

Under the firft rate or clafs in the above table I fhould put that part of the channel of moft great rivers which

is

is in extenſive plains next the ſea; with regard to *the declivity thereof alone*, but not at all with regard to the velocity of the current there, which is often very great from the compreſſion and impulſion of the upper waters, as has been repeatedly ſhewn above muſt be the caſe (N° 29. 38. 43. 72.).

Second rate or claſs. Moſt artificial canals in flat countries, made for the uſe of navigation; eſpecially thoſe in the Dutch and Auſtrian Netherlands.

Third rate or claſs. The river Trent; the Scheld and the Lys below Ghent; the Iſere and the Iprelee below Fort Knock in Flanders; many rivers in the territories of Bologna and Ferrara in Italy; the river Meander in Natolia; and innumerable others in flat countries.

Fourth claſs. The Thames; the Lys and the Scheld above Ghent in Flanders; the Senne, the Dyle, and the Demmer, in Brabant; the Seine and the Somme in France; the Nile and the Niger in Africa; the rivers of St. Lawrence below Lake Ontario, tne Oroonoko, the river of Amazons, and the rivers of Paraguay, in America.

Fifth claſs. The Severn and Ouſe in England; the Loire and Garonne in France; the Tagus, the Guadiana, and the Guadalquivir, in Spain; the Po and the Tiber in Italy; the Meuſe, the Rhine, and the Elbe, in Germany; the Weiſſel, the Neiſter, the Bog, and the Nieper,

in

in Poland; the Don and the Dwina in Ruffia; the Amur or Saghalien in Tartary; the Yellow and Blue Rivers in China; the rivers of Cambodia, Ava, and Ganges, in India; the Euphrates; the river Zaire in Congo; the Miffifippi.

Sixth clafs. The Rhone in France; the Ebro and Douro in Spain; the Danube; the Wolga; the Irtifch and Oby, the Jenefea and Lena, in Siberia; the river Indus; the Tigris; the Malmiftra in Cilicia.

Seventh clafs. In this clafs can only be enumerated thofe parts of rivers where they defcend from among mountains into the plain country below; as alfo fome rivers paffing through the midft of mountains.

Eighth clafs. To this clafs belong all torrents among mountains; fuch, for example, as the Bourns in the Highlands of Scotland are defcribed to be.

SECTION VI.

A general and eafy method of taking levels though largie extents of country where rivers pafs; and dfo of computing the heights of interior parts of continents above the furface of the fea.

80. After all I have faid hitherto in this effay, and particularly in the foregoing fection, what I am about to

lay

lay down under this laft head of it muft appear very plain and eafy. I am very far, however, from giving the methods I am going to propofe *for taking the levels through whole countries and continents as far as rivers extend*, as ftrictly exact; I know very well that it is next to impoffible they fhould be fo, confidering the continual variations in the declivities of rivers, and in the velocities of their currents in different parts, as alfo the impoffibility of knowing the exact length of their courfe through all their turns and windings I only give them therefore as a general and eafy method of computing the relative heights of countries without deviating much from the truth, which, perhaps, is all that may be neceffary for the confideration of the natural philofopher. At all events, they may be of fome ufe, for this end, in fo many parts of the earth through which rivers pafs, and where no barometrical obfervations, or any others whatever, for taking heights above the fea, have been, or perhaps ever will be made. They may alfo be found ufeful in taking the levels through a large extent of flat countries where regular canals and rivers pafs, and where the difference of elevation is too fmall to be obferved by the barometer, and where alfo the taking them through fo great an extent by the common methods of levelling would be much too expenfive for the purpofes required. Now in

I   this

this laſt caſe I have found, by experience, that by the method I here propoſe the difference of heights may very eaſily be found, and that very near to the truth.

, For this end it may be proper to premiſe a few neceſſary conſiderations and precautions to. be obſerved. in making uſe of the method I here propoſe. They would eaſily occur to any one who conſiders the principles whereon it is grounded; but to ſave trouble I ſhall put them down in a few words.

81. The firſt is, that a particular attention muſt be had to the quantity of water actually in the river at the time of the operation, ſo that according as the greater or leſs quantity thereof may augment or diminiſh the velocity of the current; allowance may be made conformable thereto in determining the quantity of declivity from the degree of velocity. .

2dly, Obſerving this precaution throughout the whole river, or all that part of it wherein we want to find the difference of elevations, we muſt next endeavour to determine, as near as poſſible, by the principles laid down in the laſt ſection, *all the variations of declivity from the variations of velocity within thoſe limits, and alſo the exact length and quantity of each.*

3dly, The ſame attention muſt be had in taking the difference of heights by canals, while their ſluices and,

communications are kept conftantly open, fo as to effec-
tuate a compleat natural current throughout the whole
extent thereof; for in this cafe they are no other than
rivers, and their waters follow the fame laws of motion.

4thly, But in canals which are fhut, and their waters
kept up by fluices fo as to render them nearly ftagnant,
the practice of this method will be different from what
it is in rivers and open canals: for in this cafe there cannot be allowed for the declivity of the furface of the water from fluice to fluice above one inch, or two at moft,
in each mile of length, according as there may be fewer
or more accidental drainings of the water in it (N° 74.).

Again, as it may happen, in taking the levels of countries by the means of artificial canals, that the water in
different parts may have different directions, attention
muft be had to *add* or *fubtract* refpectively the total declivity of each.

Moreover, it almoft always happens, in canals where
the fluices are fhut, that the water on the two fides of
each fluice is of a very different height, the back waters
being kept up, while the lower are run off to a certain
point; but in fluices next the fea, the tide againft the
outer gates is fometimes lower and fometimes higher
than the water in the canal above. In all thefe cafes, the
difference of height muft be exactly meafured, and the

quantity

quantity refpectively added or fubtracted in the account of the levelling.

5thly, After this it is neceffary to determine, as nearly as poffible, the length of the canals and rivers through all their turns and windings, and throughout the whole extent of country in which we want the difference of elevations. This may be done by an actual menfuration, or by the general opinion of the inhabitants of each part of the country, which, being founded upon the long and continually repeated experience of an infinity of people, will be found to differ very little from the truth, *attention being had to the quantity of their nominal meafures*; even the errors in *more* or *lefs* will nearly compenfate each other; or, finally, in great extents it may fuffice to compute them from good geographical maps.

6thly, This being done by one or the other of thefe methods, it will be eafy, from the quantity of declivity before determined for each part in particular, to find the whole quantity of declivity throughout the whole extent of country meafured, or from any one part thereof to any other along the rivers or canals in queftion, which are fuppofed to be continued without interruption from one place to the other. If to this be added the relative height of the country in each place compared with the level of the water in the part of the river or canal next to each,

we

we fhall have very nearly the difference of elevation of thofe two parts of the country. And thus the levels may be taken from the fea through any extent of country, nay even through whole continents, as far as rivers or canals extend without interruption. Cataracts themfelves, fuch as thofe in the Nile and in the river of St. Lawrence, need not hinder the operation, fince we have only to take the refpective heights from which they fall into the account as we do in common fluices, and allow for the increafe of velocity produced by them in the current of the river above and below the places where they exift.

82. Although I do not pretend to equal this method (of finding the difference of heights in countries) *for exactnefs* to the levels taken by actual menfuration, or to thofe found by a long feries of nice barometrical obfervations; yet it muft be allowed, that it is free from many inconveniencies, and accompanied with many conveniences, which the others are not. It may be eafily carried through great extents of country, where the other methods cannot be put in practice, on account of the expence or time required; and this may be done with very little trouble, and perhaps with fufficient exactnefs to anfwer all the purpofes of the natural philofopher in his confiderations on the globe we inhabit. Although the method of taking heights by barometrical

obfervations

obfervations is highly ufeful, and among mountains (where mine can be of little or no fervice) far preferable to every other hitherto difcovered; yet it will eafily be acknowledged by every one who is acquainted with what M. DE LUC [m], Sir GEORGE SHUCKBURGH, and Colonel ROY [n], have done upon this fubject, that the greateft attention to an infinity of varying circumftances, as well as the greateft nicety and exactnefs both in the inftruments and in repeated obfervations, are neceffary if we would come at the truth thereby.

Again, the method of taking the difference of heights by the quantity of declivity in rivers requires no attention to the curvature of the globe, an object (as every one knows) infinitely too confiderable to be neglected in the common method of levelling; as are alfo the great and varying refractions of the vifual rays fo near to the furface of the earth as they muft be taken in the practice of that method. The quantity of effects and of errors in the vifuals proceeding from this laft caufe muft be very different at different times, as it depends wholly on the greater or lefs denfity, on the greater or lefs quantity of vapours fufpended in the loweft part of the atmo--

(m) See his work on Barometers and Thermometers, in two vol. quarto.
(n) See the learned and curious treatifes of thefe two gentlemen in the Philofophical Tranfactions for 1777.

fphere,.

fphere, the ftate of which feldom remains long the fame. Now it is no eafy matter either to determine the quantity of thefe exactly, or to calculate the effects and errors in the vifual rays proceeding therefrom, which yet muft be done to come at the truth by the common method of levelling; whereas, in the method I propofe, no fuch confiderations are neceffary, as is evident from the nature of it.

But this is more than enough on a method fo obvious and eafy; I fhall now give a few examples of it, and thereby conclude this effay, already perhaps much too long.

83. Suppofing the length of the Scheld, between Antwerp and Ghent, following all its meanders, to be. forty meafured Englifh miles, as it is reckoned nearly to be; and fuppofing the length of the faid Scheld between Ghent and Tournay to be fifty of the fame miles; and that of the Lys from Ghent to, Commines, where it approaches neareft to the city of Ipres, forty-fix miles; it is required to know the refpective differences of elevation between all thefe places.

It may be found above (N° 78. 79.) that the river Scheld, between Ghent and Antwerp, has not above one foot declivity in each mile of its courfe; and that the

I   Scheld

Scheld and the Lys, above Ghent, have about one foot declivity in each four thousand feet of length.

According to this, the surface of the Scheld in Ghent is about forty feet higher that it is at Antwerp; and at Tournay it is sixty-six feet higher than at Ghent, and one hundred and six feet higher than at Antwerp. So also the surface of the Lys at Commines is sixty-one feet higher than at its junction with the Scheld in Ghent, and one hundred and one feet higher than the same at Antwerp. From hence it may be deduced, that the Scheld at Tournay is about five feet higher than the Lys at Commines, through twenty-five miles of interjacent country.

84. Suppose it be required to find the difference of height between the surface of the Lys at Commines and the surface of the canal at Ipres which falls into the sea at Nieuport on the Coast of Flanders. The distance between Ipres and Commines is nearly seven measured miles, through which there is no communication by water; but there is one a great way round, which therefore, for the purpose required, must be followed through all the differences of elevation comprized therein, *viz.*

Descent

Defcent towards the fea at Nieuport *on the Coaft of* Flanders.

Feet.

Total declivity of the Lys from Commines to Ghent, 46 miles,   -   -   -   -   -   61

Declivity of the canal from Ghent to Bruges, when the fluices are fhut,   -   -   -   1

Difference of height of the water in the aforefaid canal above that in the canal from Bruges to Oftend, which two communicate together by fluices,   -   -   -   -   8

Declivity from Bruges to Plafchendahl where the canal of Nieuport joins that of Oftend,   -   6

Total declivity from Plafchendahl to Nieuport, in-cluding the difference of furfaces in two interme-diate fluices,   -   -   -   -   8

Total declivity, all one way, from Commines to Nieuport, about 95 miles,   -   -   84

Afcent

too far into the regions of conjecture; but as such mistakes as these are no ways prejudicial to my fellow creatures, to whom I wish to be useful, and as they may give occasion for others to rectify them, and so lead them to a subject which otherwise, perhaps, they might never have attended to, I shall hope for indulgence from all those who wish well to humanity and to useful knowledge.

from what was found by actual levels made from the Lys to Ipres by the French engineers during the time that LEWIS the XIVth was mafter of the country, when there were propofals for opening a canal from the one to the other.

85. I fhall venture to carry my conjectures ftill farther, and grounding them upon the principles laid down above (N° 78. 79.) I fhall take a general view of the elevations of continents along the courfe of the principal rivers in the known world. I cannot, however, repeat too often, that I give this as a matter of mere conjecture and curiofity. It has not been, nor ever will be, in my power, or in that of any other particular perfon whatfoever, to follow the courfes of all the rivers mentioned in the enfuing table from their mouths to their fources. All that can poffibly be done on this head, is to examine the relations of voyagers and geographers concerning each river as far as it is known, and to reduce it by that means within the compafs of the hydrometrical principles laid down in this effay. This is what I have done as far as I could; and therefore, allowing that I make great miftakes therein, yet I do not think that I merit much blame on that account, as I have done what I was able to do. If I am blame-worthy, it is for having launched out

Afcent from the fea at Nieuport *to the city of* Ipres.

Feet.

Difference of height which the river from Nieuport
to Ipres has above the canal from Nieuport to
Plafchendahl, taken crofs the harbour of Nieuport, by means of the tides which come up againft
the outer gates of the fluices next the fea on each,   3

Declivity of the river Iprelee from Nieuport to
Boefinghe,   \`   -   -   -   11

Difference of level of the water above and below the
fluice of Boefinghe,   -   - $22\frac{1}{3}$

Declivity in the canal from Boefinghe to Ipres,   - $\frac{1}{3}$
___

Total afcent, all one way, from Nieuport to Ipres,   37
___

Now $84-37=47$ feet for the difference of height
which the furface of the Lys at Commines has above
the furface of the canal at Ipres. I have made ufe of the
above example preferably to any others, as it is very
complicated, and becaufe the quantities of declivity
which I have put down are not arbitrary; and, moreover,
becaufe I had the good fortune to find, fome years after I
had taken thofe meafures, that I only differed two feet

86. A table of the elevation of countries above the surface of the sea, at each 100 miles of length up the course of the principal rivers in the world, as far as they extend; by computation from the principles laid down in this treatise.

Feet of elevation in 100 miles.	Feet of elevation in 200 miles.	Feet of elevation in 300 miles.	Feet of elevation in 400 miles.	Feet of elevation in 500 miles.	Feet of elevation in 600 miles.	Feet of elevation in 700 miles.	Feet of elevation in 800 miles.	Feet of elevation in 900 miles.	Feet of elevation in 1000 miles.	Names of Rivers, And quantities whereby the length of their course is to be diminished, to have the distance from their mouths in a direct line.
100	210	330	470	630	820	1040	1300	1600	1950	The river Trent, the Meander, and many others of the same kind, which are seldom of great extent: in these one-third of the course may be allowed for its deviations from a right line.
150	310	480	670	880	1120	1400	1770	2160	2620	The river Thames; the Seine and the Somme in France; the Nile and the Niger in Africa; the river St. Laurence, the Oroonoko, the river of Amazons, the rivers of Paraguay: in these about one-fourth of the length of course may be allowed for turns and windings in it.
220	450	700	980	1290	1640	2040	2500	3030	3630	The Severn; the Loire and Garonne in France; the Tagus, Guadiana and Guadalquivir in Spain; the Po and the Tyber in Italy; the Meuse, Rhine and Elbe in Germany; the Weixel, Neister, Bog and Nieper in Poland; the Don and Dwina in Ruffia; the Amur in Tartary; the Hoang-ho-keou and Yang-tfe Kiang-kcou in China; the rivers of Cambodia, Ava and Ganges in India; the Euphrates; the Zaire in Congo; the Miffiffippi: in these may be allowed about one-fifth of the length of the course for turns and windings in it.
300	650	1050	1520	2070	2720	3500	4440	5570	6920	The Rhône in France; the Ebro and Douro in Spain; the Danube; the Wolga; the Irtifch, Oby, Jeneſca and Lena in Siberia; the Malmiftra in Cilicia; the Tigris; the Indus. The course of these rapid rivers is usually very straight, and there cannot be above one-fixth of the length thereof allowed for deviations from a right line.

www.ingramcontent.com/pod-product-compliance
Lightning Source LLC
Chambersburg PA
CBHW021828190326
41518CB00007B/782